Trust and Virtual Worlds

Steve Jones
General Editor

Vol. 63

PETER LANG
New York • Washington, D.C./Baltimore • Bern
Frankfurt • Berlin • Brussels • Vienna • Oxford

Trust and Virtual Worlds

CONTEMPORARY PERSPECTIVES

EDITED BY CHARLES ESS & MAY THORSETH

PETER LANG
New York • Washington, D.C./Baltimore • Bern
Frankfurt • Berlin • Brussels • Vienna • Oxford

Library of Congress Cataloging-in-Publication Data

Trust and virtual worlds: contemporary perspectives /
edited by Charles Ess, May Thorseth.
p. cm. — (Digital formations; v. 63)
Includes bibliographical references and index.
1. Virtual reality—Social aspects. 2. Cyberspace—Social aspects.
3. Trust. 4. Telematics—Social aspects. I. Ess, Charles. II. Thorseth, May.
HM851.T78 303.48'34—dc22 2011004202
ISBN 978-1-4331-0923-2 (hardcover)
ISBN 978-1-4331-0922-5 (paperback)
ISSN 1526-3169

Bibliographic information published by **Die Deutsche Nationalbibliothek.**
Die Deutsche Nationalbibliothek lists this publication in the "Deutsche Nationalbibliografie"; detailed bibliographic data is available on the Internet at http://dnb.d-nb.de/.

The paper in this book meets the guidelines for permanence and durability of the Committee on Production Guidelines for Book Longevity of the Council of Library Resources.

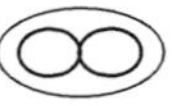

29 Broadway, 18th floor, New York, NY 10006
www.peterlang.com

Printed in the United States of America

Contents

Introduction

CHARLES ESS
MAY THORSETH

Origins

On a popular level, the term "virtual" appears to have lost some of its novelty and zing. We take this to signal that our fascination with and enthusiasms for "virtual reality," "virtual worlds," even "virtual communities" have calmed a bit, a least in comparison with the early 1990s, as these enthusiasms and fascination were captured in and further catalyzed by the seminal works of Howard Rheingold (1993), Michael Heim (1993) and Sherry Turkle (1995). Indeed, at least on the level of popular culture, the once unquestionable salience of "the virtual," "virtual reality," and so forth was relatively short-lived. So, for example, by 1996, the character Angel in the musical "Rent" could poke bitter fun at M.I.T.'s virtual reality projects, accusing virtual reality of distracting us from another pressing reality, especially for embodied beings—namely, AIDS.[1]

In ways that we explore in this volume, at least some portion of that early enthusiasm rested on a presumed, hard-and-fast dichotomy between "the virtual," as shorthand for the sorts of communicative interactions made possible by computer-mediated communication (CMC), and "the real"—where the latter mapped onto a material world whose darker sides include sexual objectification and victimization, vast economic disparities, political oppression of every stripe, and, indeed, mortality itself. Given such a dualism, it is little wonder that the multiple promises of "liberation in cyberspace"—including virtual worlds and virtual communities seemingly free of these otherwise inescapable material realities—evoked passionate enthusiasm and zealous defenders. For better and for worse, however, with some important exceptions, the boundaries and thus the presumed oppositions between the virtual and the real as pegged to the online and the offline have softened, if not largely dissolved. Broadly, what Barry Wellman calls the second age of internet studies begins somewhere around 1998—an age marked by greater emphasis on empirical data and thereby greater insight into the ways in which all things internet more and more seamlessly interweave and mesh with the multiple nooks and crannies of the material world (2010, p. 19). In particular, by this same year, Pierre Lévy critiqued earlier dualistic understandings that opposed "the virtual" and "the real," arguing instead for a more complementary relationship between the two (1998).

Whatever losses and disappointments these shifts may portend for earlier utopian hopes and dreams—for philosophers, this cooling down from often intense passions surrounding assumptions and views of "the virtual" is beneficent. As we (and certainly others) know, Hegel's Owl of Minerva—the icon of philosophical reflection and understanding—flies only at dusk. That is, despite the insistence of our technologically driven age to make everything available to everyone instantaneously—many good things take time: this is especially true for careful and informed philosophical reflection. And so it seemed to us in late 2008 that the time was propitious for systematic philosophical reflection and debate on "the virtual." In particular, the matters of trust and deliberation online as refracted across the differences between online and offline environments had become an increasingly significant focus of philosophical attention: indeed, a number of well-developed philosophical analyses and theoretical approaches to the virtual and virtual worlds, and to trust and deliberation online, had emerged in the literature. But to our knowledge, no one had yet managed to bring these diverse frameworks and analyses together for the sake of a comparative debate and discussion—much less, in order to test and refine the more well-developed and established theories in light of some of the most recent technological developments and philosophical analyses.

These then became the large goals of a workshop we organized for both faculty and graduate students, entitled "Philosophy of Virtuality: Deliberation, Trust, Offence and Virtues," held on the campus of the Norwegian University of Science and Technology (NTNU) on March 9–13, 2009. Our gathering was made possible by generous financial and logistical support from the Programme for Applied Ethics, housed in the Philosophy Department, NTNU. In order to address our larger goals, we sought to bring together both senior and more junior philosophers in this workshop, all of whom were enjoined to address diverse topics of virtuality from philosophical and ethical perspectives, with specific attention to such focal points as deliberation and trust online, among others.

The chapters collected here are drawn from presentations first developed for the workshop—and this genesis defines both their foci and limits. To begin with, our shared groundwork is squarely philosophical. This means that readers will not find here much discussion of or reference to the extensive treatments of virtuality found, say, in computer science or communication studies. By the same token, our focus on trust is likewise squarely directed at how philosophers have taken up trust in both offline and online contexts. Again, this focus largely excludes extensive and important work on trust in the fields of economics, business, and e-commerce. (The exception here is the chapter by Mariarosaria Taddeo, which draws fruitfully on the literature of e-commerce.) To be sure, a more complete approach to trust and virtual worlds would require extensive attention to these fields. But by restricting our focus to the philosophical, we believe that the chapters collected here thereby bring to the foreground a range of arguments and analyses that are frequently missing in other approaches and literatures. Indeed, at the risk of immodesty, as philosophers we believe these insights and analyses may prove crucial, if not in some ways foundational, to further discussion and debate within and across the multiple fields that take up notions of trust and the virtual.

Trust, the virtual, and Web 2.0: Organization of the Volume

As Charles Ess, Bjørn Myskja, and others collected here make clear, *trust* is of compelling interest vis-à-vis virtual environments for two specific reasons. First of all, without trust, there is no human society and there is thus no good life. Secondly, as Myskja has previously pointed out, the Danish philosopher and theologian Knut Løgstrup has highlighted the importance of our *embodied* co-presence with one another as the primary condition for our overcoming an initial wariness and mistrust (Myskja, 2008, p. 214, with reference to Løgstrup, 1956, p. 22f.). But especially in the early forms of online encounters with one another, including the MUDs and MOOs that Rheingold, Turkle, and others took as exemplars of virtual communities, the only medium of communication was *text*—i.e., narrative descriptions without the benefit of sound or vision (whether as a still photograph or a video). As Myskja, among others, elaborates here, numerous scholars in information ethics have argued that such environments, as essentially disembodied, in these ways, thus raise central challenges for establishing and sustaining trust online.[2]

But of course, the increasing interweaving between the virtual and real in contemporary life is facilitated by technological and infrastructure developments that enable more and more use of sound and video, whether in the form of an avatar in *Second Life*, a YouTube video, or a videoconference call that uses simple webcams to simultaneously carry live video alongside sound. Our broad question—explored here through nine chapters—is how far trust, as central to human society and the good life, is both challenged and fostered by the virtual environments made possible especially over the past five years in affiliation of the rubric of "Web 2.0."

We explore this question as follows. "Section I: Historical and Conceptual Perspectives" is made up of contributions from Charles Ess, Marianne Richter, and Johnny Søraker. These approach virtuality and trust from the perspectives of CMC, information ethics, and phenomenology (Ess), philosophical and linguistic analysis (Richter), and philosophical and empirical analysis (Søraker). Collectively, these contributions provide basic historical and philosophical frameworks that highlight fundamental difficulties surrounding notions of virtuality and establishing trust in online environments, coupled with some initial resolutions to these problems.

Chapters by Mariarosaria Taddeo, John Weckert, Annamaria Carusi, and Bjørn Myskja constitute "Section II: Philosophical perspectives on trust in online environments." Taddeo and Weckert develop fine-grained philosophical analyses of different forms of trust and their possible realization in online environments. Carusi, drawing in part on phenomenology, characterizes our experience as essentially intersubjective, one in which trust is already in play to at least a minimal degree: hence the "virtual"/"real" divide is less problematic for trust online than we might initially think. Myskja argues that the (usually) greater potential for deception regarding one's identity in online environments can actually work in virtuous, not simply vicious ways. Both Carusi and Myskja appeal to Kant's philosophy in key ways, highlighting his importance for these explorations. Finally, "Section III: Applications/implications," explores specific domains in which trust and virtuality come into play in central ways. Litska Strikwerda explores philosophical and legal arguments as to whether virtual child pornography—as *entirely* virtual—nonetheless can be legitimately criminalized. May Thorseth examines how virtual environments may foster real-world democratic processes.

Section I: Historical and conceptual perspectives

We start, then, with **Charles Ess**'s "Self, Community, and Ethics in Digital Mediatized Worlds." This chapter provides a historically oriented overview of how the key notions of "virtual" and "real" were conceptually mapped out, beginning with the relatively early attention to "cyberspace" in the 1980s, including William Gibson's science-fiction novel *Neuromancer* (1984). *Neuromancer* appears to have defined much of both popular and scholarly understandings of virtual worlds and online experiences more broadly, especially as it cast these in sharp opposition to "the real" as correlated with our embodied existence as material beings offline. These dualisms thereby raise a central problem for the establishment of *trust* in virtual environments. As Ess summarizes representative understandings and trust in philosophy, it becomes clear that trust—at least initially—requires our embodied co-presence with one another. But of course, such embodied co-presence is precisely missing in at least those instantiations of virtuality that celebrate instead (and in some cases, at least, with good reason) the absence of the body.

Such dualisms can be seen specifically at work in 1990s' enthusiasms for virtual communities, as Ess documents—but further extend to other instantiations of the online and the virtual, most importantly with regard to postmodernist and post-structuralist conceptions of identity and identity-play made possible in online environments such as MUDs and MOOs. With both the ongoing diffusion and expanding uses of the internet and the web, on the one hand, and increasingly sophisticated CMC research, on the other, these dualisms are gradually replaced, however, by the increasing reality and recognition that the offline and the online, and thereby (most) instantiations of the real and the virtual are seamlessly interwoven. Ess goes on to show how parallel developments in phenomenology and neuroscience reinforce these anti-Cartesian turns in our use and understanding of the virtual and the online—thereby helping to overcome the early dualisms that rendered trust in these environments fundamentally problematic.

These developments point to the possibilities of more fruitful conjunctions of trust and virtuality. But as **Marianne Richter** shows in her "'Virtual Reality' and 'Virtual Actuality': Remarks on the Use of Technical Terms in Philosophy of Virtuality," there are additional conceptual and linguistic hurdles surrounding our key terms *reality, actuality* and *virtuality* that, at least from a philosophical perspective, must first be cleared. Richter presents a detailed exploration of the origins and characteristic uses of these key terms, with a specific interest in attempting to determine what "virtual reality" and "virtual actuality" can be understood to mean. Her analyses make painfully clear, however, that there is a great deal of philosophical confusion and linguistic ambiguity marking characteristic uses of these terms.

The central philosophical problem comes through the fundamental distinction between "modal" and "generic" terms. As Richter explains,

> generic terms [are] super-ordinate concepts that allow for the distinction between sub-concepts (e.g., "elephant" and "dolphin" can be subsumed under the generic term "animal"). Modal terms refer to a different usage of terms in the way that they denote modalities of facts or propositions. Basic modal terms are "necessary," "possible" and "actual." (Richter, p. 42)

Richter goes on to point out that if we want to remain conceptually clear, in light of this distinction we can use a term either as a generic term or as a modal term—but not both. But this is precisely what we do when we use "virtual" as both a generic term (i.e., referring to a class of instantiations that can be categorized as subcategories, e.g., virtual communities, virtual romances—or, to use one of Johnny Søraker's examples, virtual memory) and as a modal term, as when "virtual *reality*" and "virtual *actuality*" thereby point to specific modes of being.

Richter acknowledges that especially when dealing with new and emerging technologies, it is out of place to demand perfect conceptual and linguistic coherency. At the same time, however, she rightly points out that especially from a philosophical perspective, "*coherent* use of technical terms seems to be justifiable with regard to a pragmatic argument: in terms of inter-subjective exchange and development of ideas, it certainly is helpful to refer to a common conceptual and terminological basis" (Richter, p. 32) Otherwise, to state it more bluntly, the risk is that we may literally not know what we are talking about.

And so we turn to **Johnny Søraker**'s "Virtual Entities, Environments, Worlds and Reality: Suggested Definitions and Taxonomy." Søraker reiterates Richter's point: "... there is hardly any consensus on what these terms ['virtuality,' 'virtual reality,' etc.] should refer to, nor what their defining characteristics are" (Søraker, p. 44). As he points out, this lack of precision becomes especially problematic precisely when we seek to distinguish between "the virtual" and "the real"—as in the example of "virtual communities" or "virtual trust."

Søraker seeks to resolve these confusions by first reviewing some of the prevailing definitions of the virtual, so as to build towards a defensible taxonomy of virtuality. This taxonomy takes on board Richter's distinction between generic and modal terms, and starts by defining "virtual" as a generic term, i.e., one that refers to a number of sub-categories. For Søraker, "virtuality" includes as necessary (but not sufficient) conditions both interactivity and dependence upon computer simulation. Søraker then carefully distinguishes between "virtual environments," "virtual worlds," and "virtual reality," as these are marked (respectively) by "indexicality" (a term referring to a complex of requirements, primarily that of enabling a sense of location in a three-dimensional space), multi-access (i.e., the possibility of more than one person participating in a venue simultaneously), and first-person view (most familiar from high-tech virtual reality systems employing head-mounted displays). These distinctions issue in a taxonomy composed of "virtuality" as a (properly) generic term. This includes "virtual environments" (which require indexicality) and which in turn has two sub-classes: "virtual reality" (which additionally requires a first-person view) and "virtual worlds" (which additionally require multi-access). We can note specifically that "virtual community" on this taxonomy is defined in terms of "communication between multiple users that is computer mediated and interactive, but not necessarily taking place in a three-dimensional environment" (p. 61). This is to say that a "virtual community" of a text-based sort, such as the early MOOs and MUDS, can exist as a sub-class of virtual worlds—but thereby remains distinct from the sorts of interactions, e.g., via avatars in a 3-dimensional space such as *Second Life*, which count rather as virtual worlds. (See his Figure 2 and Table 2 for helpful summaries, p. 65).

In developing his taxonomy, Søraker draws directly on Husserl's phenomenological account of how we experience the world to develop these definitions—contributing thereby to the larger phenomenological thread that runs throughout our volume. More broadly, Søraker's larger goal is to contribute to the conceptual clarity needed in discussing various instantiations of virtuality, especially vis-à-vis important ethical and social considerations. As we have seen, for example, and as he reiterates by way of Howard Rheingold (2000, p. 177), the absence of the body—and thereby of bodily gestures and facial expressions—makes *trust* in virtual *communities* (at least of the text-based sort) problematic. But where such communities exist, by contrast, in what Søraker defines as a virtual world as a subclass of virtual environments—i.e., as including precisely some form of presence in a three-dimensional space—then the question of trust becomes a different one from that evoked by virtual communities *per se*.

Having reviewed some of the relevant historical background and, we think, overcome significant conceptual and linguistic difficulties attaching to earlier work on virtuality, we can now turn to more specific philosophical analyses.

Section II: Philosophical perspectives on trust in online environments

In her "The Role of e-Trust in Distributed Artificial Systems," **Mariarosaria Taddeo** analyzes a specific form of trust—namely, "e-trust" as a precisely defined set of relations between Artificial Agents (AAs). Taddeo's account provides a first example of what philosophers characterize as *rationalistic* accounts of trust, i.e., accounts that emphasize how far trust depends upon our having good *reasons* to trust another person or agent. Specifically, Taddeo follows here "a Kantian regulative ideal of a rational agent," one that is "thereby able to choose the best option for themselves, given a specific scenario and a goal to achieve" (Taddeo, p. 76). Her account draws on rational choice theory, and is distinctive, as she defines e-trust not as a first-order relationship between two AAs, but rather as a second-order property of their relations.

For Taddeo, a chief advantage of this approach is that developing an account of trust for AAs is simpler than developing an account of trust for humans. Taddeo points out that trust among human agents is more complex because it implicates many more factors (such as economic and psychological factors) and must take place within a much greater range of more complex circumstances than those defined for AAs. Specifically, Taddeo notes that AAs are "not endowed with mental states, feelings or emotions" (p. 76). We will immediately explore these important contrasts between trust among humans and trust between AAs in John Weckert's chapter. But it is important to emphasize here that a definitive advantage of Taddeo's definition and her approach more broadly is that it can be instantiated in real-world computational devices, ranging from stock-market trading programs to predator drones. In this way, e-trust provides an empirically testable definition of trust which thereby allows us, as Taddeo suggests, to follow the model of a good scientist who seeks to test her hypothesis through experiment.

John Weckert's "Trusting Software Agents" begins by distinguishing trust from what Weckert calls "mere reliance"—namely, reliance upon artifacts. That is, trust between human beings seems to entail the operation of *choice*: that is, we *choose* to trust people—who in turn can choose whether or not to behave in the ways we entrust them to. By contrast, it is more precise to say that we *rely* on, in Weckert's examples, artifacts such as bicycles or ladders: but unlike human agents, these artifacts have no choice in whether or not they will live up to our expectations. Weckert further presents trust as a form of "seeing as"—in contrast with (though not necessarily in conflict with) the strict sort of rational choice theory Taddeo introduces for us. Comparing this to Kuhn's analysis of the paradigms that drive normal science, Weckert suggests that trust is something like a hermeneutical framework (our term) that shapes our experience and perception.

This leads Weckert to define trust as a kind of reliance—but not "mere reliance." That is, "we will say that if *A* trusts *B*, then *A* relies on *B*'s disposition to behave favourably toward *A*" (Weckert, p. 91). For Weckert, "disposition" here includes the central component of choice or autonomy. At the same time, this definition highlights three additional and important features of trust. One, trust involves vulnerability and risk (as Myskja will reiterate by way of Løgstrup). Two, in part because it derives from human choices, trust is thereby *not* an epistemic certainty. This means, rather, that trust is not entirely a matter of decisive reasons. As well, finally, there are noncognitive and affective dimensions to trust—in Weckert's example, a young child will (likely) trust his or her parents, while not doing so simply as a matter of having toted up the reasons pro and con as to why these parents can be trusted.

Given the central role of choice and thereby autonomy in human experiences of trust, the central question for Weckert is how far we can be meaningfully said to trust software agents. On the one hand, it would seem—as Taddeo emphasizes—that such agents do not possess the sort of autonomy we take human beings to have. Against this, however, Weckert argues that software agents do possess a kind of autonomy that is sufficiently close to human autonomy as to argue, "that autonomous software agents can be trusted in essentially the same sense in which humans can be trusted" (Weckert, p. 100).

A key issue raised between Taddeo and Weckert—and one we will return to in our conclusion—is thus: precisely how far are human beings, especially on a Kantian approach, defined by a moral autonomy that may not be within the capacity of autonomous agents?

Annamaria Carusi, in her "Trust in the Virtual/Physical Interworld," approaches the matter of trust in online environments from the phenomenological tradition. As the terms in her title—"virtual/physical" and "interworld" (the latter deriving from Merleau-Ponty) suggest—Carusi argues against a strong dichotomy between the virtual and the physical. She further criticizes exclusively rationalist accounts of trust, insofar as these depend upon a comparatively external approach: by contrast, she argues that trust can only emerge out of an "intersubjectively shared world," one that further depends upon "a common perceptual as well as conceptual system" (p. 117).

Part of Carusi's critique of strictly rationalist accounts echoes the well-known problem that we can call the vicious circle of security. Briefly, given uncertainty as to whether or not we can trust another, especially in online environments, we may be inclined to build ever stronger and more elaborate systems of regulation and security mechanisms. As May Thorseth has pointed out elsewhere, however, this approach ultimately works to *undermine* trust: "...such a strategy is to bite one's own tail: where the control mechanism has replaced an initial trust between citizens, there is little reason to believe that the trust can be regained through increased control" (2008, p. 130). Similarly, Carusi argues that strictly rationalist accounts can lead to demands for reason-giving that unfortunately fall under Sartre's notion of bad faith (Carusi, p. 109). The counter to this is precisely to recognize where trust already exists in our intersubjective world: this allows for a kind of moral bootstrapping or virtuous circle in which initial trust (to echo Weckert, above)—specifically, acting *as if* others are trustworthy (anticipating Myskja, below)—leads to greater trust in online environments. Carusi exploits these insights, along with her extensive experience in e-Science projects that require such trust, to develop four useful guidelines for fostering trust in online environments.

As we have now seen in a variety of ways, a central challenge to trust in online environments is the multiple ways in which these environments allow us to easily disguise or hide our embodied identities. **Bjørn Myskja**'s approach to this problem in his "Trust, Lies and Virtuality" is novel and distinctive insofar as he argues that this potential for vice can also work as a potential for virtue. Even more strikingly, he does so by way of Kant's ethical philosophy, beginning with its radical injunction against lying as a premier example of how the Categorical Imperative (CI) functions. As is well known, the CI in early Kant requires us never to lie, even when there might be clear and otherwise legitimate benefits to do so, such as saving an innocent life. In this direction, Mysjka's use of Kant highlights a crucial connection between truthfulness and trust: "It is our basic moral duty according to the Kantian moral philosophy to act in accordance with the trust others place in us, that is, to act in a way that make us deserving of trust." The surprise, as Myskja will argue, is that "[w]e can be trustful in this sense, while allowing for some evasions of telling the truth" (p. 124).

To make this case, Myskja first draws on Kant's later (and, we might agree, more realistic) moral philosophy to recognize that at least in certain ways, deception can play a virtuous role. Briefly, we pretend—in Kant's phrase, we act *as if*—we are better than we are. And even though we know others are doing the same, Kant argues that acting *as if* we are better in these ways actually leads to our becoming and *being* better over time. This then allows Myskja to take the potential for deception online—ordinarily conceived as a major obstacle to trust in particular and virtue more broadly—and highlight how this potential can thus play a *positive* role. Most briefly, the relatively greater freedom for such deception in online contexts thereby offers greater positive opportunity for self-development (as well as self-expression).

Echoing Carusi's description of a kind of moral bootstrapping that comes by presuming trust, Myskja's account thus argues by treating others *as if* they are trustworthy can lead them (and us) to in fact behave in more trustworthy ways over time. In this way, Myskja provides us with a strong line of argument that highlights how deception online, *contra* prevailing views that see in such deception a central obstacle to trust, can instead work in a virtuous direction—just as it does in our real-world lives. His account thus complements Carusi's defense of trust in online environments, insofar as Myskja's does so while maintaining at least some distinction between offline and online experiences and behaviors.

Section III: Applications/implications

Here we turn to two specific analyses of online behavior that help us explore in greater detail still relevant differences between online and offline environments and their ethical, legal, and political implications.

We begin with **Litska Strikwerda**'s "Virtual Child Pornography: Why Images Do Harm from a Moral Perspective." Strikwerda provides a careful and detailed analysis of the multiple legal arguments surrounding the production, distribution, and possession of virtual child pornography. She thereby gives us a highly focused and specific test case study that allows us to explore the more general and/or theoretical questions clustering about prevailing conceptions of "virtual" and "real" with both a very strong empirical foundation (i.e., as she provides us with an overview of relevant effects studies and related arguments) and legal perspectives. This test case is of especial significance in our larger project as *virtual* child pornography brings us to the limit of any distinction we might want to make between "the virtual" and the "real"—i.e., such materials are entirely independent of any real children in the material/physical world.

Strikwerda examines three main approaches to criminalizing child pornography, beginning with John Stuart Mill's harm principle as definitive of liberalism. Briefly, the harm principle is rejected insofar as the production and consumption of *virtual* child pornography appears to be a "victimless crime," i.e., one that does not involve direct harm to children. Strikwerda then turns to the paternalistic approach, one that highlights the legitimacy of society prohibiting behaviors that lead to harmful actions. In this case, the question is whether the production and consumption of virtual child pornography works to encourage pedophiles to encourage real children to engage in sex. Strikwerda again finds that the available evidence, however, does not conclusively demonstrate such a causal link, and hence the paternalism approach also fails (at least for the time being).

Strikwerda then turns to virtue ethics. A chief advantage of this approach is that, in contrast with liberalism and paternalism, "it is not necessary to prove a causal link between virtual child pornographic images and actual instances of abuse to consider them as harmful" (p. 151). Rather, a strong argument can be made against the consumption of pornography as leading to a non-virtuous mentality, i.e., one that eroticizes the domination of women and children. In particular, Strikwerda takes up the work of Sara Ruddick (1975), who uses phenomenological analyses to define "complete sex" as interwoven with a number of important virtues, including respect for persons and equality. On Ruddick's showing, however, complete sex requires *embodiment* of a specific sort: this allows Ruddick to connect complete sex with the specifically Kantian requirement to never treat other *persons* as means to our own ends (e.g., sex objects), but always as ends (i.e., unique and free beings). Strikwerda further points to empirical research that strongly links pornography consumption with distinctively non-egalitarian attitudes towards women and sex. Strikwerda thus develops a strong ethical critique of virtual pornography: such material by definition depicts "incomplete [because disembodied], non-reciprocal sex acts. They thereby violate a sexual mentality based on the equality norm" (p. 156).

May Thorseth examines the potential for online virtual environments to contribute to the enhancement of public reason, based on the ideals embedded in deliberative democracy. As a specific instance of the central challenge to trust online that we examine here (i.e., how far trust, as closely interwoven with embodied co-presence, may be established in online, especially more disembodied environments), Thorseth examines whether the proximity offered in virtual contexts is adequate for establishing the kind of trust that is required for deliberation.

Generally, deliberative democratic ideals do not necessarily seem to require embodied presence in order to obtain proximity. Specifically, however, so-called aesthetic democrats argue that embodied presence is essential to establishing the kind of emotive and passionate relationship that is needed for democratic attachment. Because passions and spontaneous associations are considered to be flattened in online communication, the virtual environments will not help improve broadened thinking, which is basic to democracy, they hold. Against this, Thorseth argues that there is no proof that emotive and passionate aspects of communication are lacking in virtual contexts, not even proximity and presence. Particularly younger and savvy users of virtual environments seem to experience communication in these contexts as no less emotional and passionate than in offline environments. Passion and proximity seem to be vital to broadened thinking, and the virtual contexts are becoming ever better at conveying embodied presence, according to Thorseth.

A main hypothesis in her chapter is that more trust will bring about more broadened thinking. Trust is here analysed as an action-based concept, in terms of "keeping one's guard down" (Grimen, 2010). The basic idea is that trust is expressed by way of not taking precautions. This concept of trust, when applied to online, virtual contexts in particular, is demonstrated by the willingness by the interlocutors to continue choosing to engage in comunicative interaction. However, whether this way of expressing trust is a sign of willingness to transcend the contingent limitations of one's own experiences has no conclusive answer.

Finally, Thorseth's analysis draws upon several of the other contributions to this volume, thereby emphasizing strong connections to both the Kantian and her - meneutical approaches running through the chapters collected here.

Concluding summaries: Trust, duties, and virtues in virtual environments

Taken together, these contributions thus help us develop first of all a rich and comprehensive account of *trust* and its multiple dimensions. This account then stands as a useful starting point for further critical reflection on how far (and if so, under what specific conditions) trust may emerge and function in contemporary instantiations of "the virtual"—where, taking up Søraker's taxonomy, we now have an equally substantive and refined account of what "the virtual" refers to in well-defined terms and specific categories and their subclasses.

Broadly, it is helpful to begin with rationalistic accounts of trust—including the one developed in exquisite detail by Taddeo, especially as this account can be instantiated on machines that operate in the material world, thereby allowing us to test the account. At the same time, however, as Taddeo notes, the capacity of such an account of trust to succeed in *praxis* depends precisely on first separating out the domain of e-trust among Artificial Agents from the more complex domains in which trust appears to operate among human beings. As we have seen, a key dimension of this complexity is how far human beings may possess an *autonomy*—especially in a Kantian sense—that cannot be fully replicated by computational devices. This issue is debated, in effect, between Taddeo and Weckert—and, nicely enough, further illuminated as we will see below, through the Kantian "red thread" (*rote Faden*) that runs through these chapters as well. Moreover, trust emerges in Weckert's account to include the noncognitive and the affective—e.g., the trust children have in their parents—which further highlights trust as an epistemic uncertainty coupled with our basic human condition of vulnerability and dependency upon others. This means, specifically, that trust involves *risk*. As Weckert has developed previously (2005), trust is nonetheless *robust*—i.e., it seems to be at least an initial given in human experience, a characterization that Carusi reinforces as her phenomenological approach highlights the pre-existence of a shared "interworld" or intersubjectivity out of which trust emerges. Carusi thereby highlights trust as involving a virtuous circle that she further describes as a kind of "moral bootstrapping": some element of trust must already pre-exist in our relationships with one another before further trust can be built. This notion of moral bootstrapping coheres well with Weckert's account of trust as "seeing as," in which trust works as a kind of interpretive framework that, as Weckert points out, functions like a Kuhnian paradigm for normal science. Such an interpretative framework is hard to call into question, much less fundamentally deconstruct. This characterization of trust, finally, directly parallels Myskja's use of Kant's notion of seeing others *as if* they are trustworthy. Specifically, what Myskja describes as the resulting virtuous circle—i.e., my seeing others as if they are trustworthy may encourage them to become more so—thus reinforces Carusi's description of trust as a virtuous circle that begins with a kind of moral bootstrapping.

Further, this summative account of trust can now be seen to be rooted in especially three strong philosophical traditions—namely, Kantian philosophy, phenomenology, and virtue ethics. These "red threads" are explicitly conjoined in

Carusi's and Strickwerda's chapters in ways that echo and reinforce additional work in these domains.

To begin with, phenomenological approaches are introduced by Ess as a philosophical development employed over the past few decades to highlight the crucial importance of *embodiment* against efforts rooted in Cartesian (if not Augustinian) dualisms that would have us radically separate mind from body. Ess shows that these phenomenological approaches were foundational to important critiques of "the virtual" as radically separate and opposed to "the real"—including Borgmann's critique of a bodiless cyberspace and Dreyfus' critique of online and thereby (at the time) largely disembodied education.[3] To recall Borgmann specifically:

> The human body with all its heaviness and frailty marks the origin of the coordinate space we inhabit. Just as in taking the measure of the universe this original point of our existence is unsurpassable, so in venturing beyond reality the standpoint of our body remains the inescapable pivot. (1999, p. 190)

Søraker likewise draws on the phenomenological tradition, as represented both by Borgmann and Husserl. Specifically, Søraker takes up Husserl's account of how our bodily actions, as immediately correlated with changes in our perceptions, thereby create for us our sense of presence in the world from our own, unique perspective—what Borgmann calls "the inescapable pivot"—as helping to define virtual *reality*. Such virtual *reality* closely mirrors our experience as embodied persons more broadly, precisely because it mimics this most foundational sense of unique, first-person perspective as defining *my* experience. As Søraker puts it:

> [B]eing in a reality requires having a first-person view, and it is impossible to have more than one first-person view. That is, you cannot participate in multiple virtual realities simultaneously any more than you can be in more than one spatiotemporal place simultaneously in physical reality. You can have as many avatars in as many virtual worlds or environments as you like, but as soon as you experience the virtual reality "through your own eyes" you can only be in one virtual reality at a time. (p. 63)

Or, to anticipate the Kantian thread—our experience of the world must always be accompanied by the "I think," i.e., precisely this sense of a unitary presence, located in a particular place: or, as Borgmann has it, "the origin of the coordinate space we inhabit" as inextricably rooted in a specific body.

Moreover, in her account of an intersubjective "interworld" (again, drawing thereby on Merleau-Ponty), Carusi thus reinforces both the emphases of phenomenology and the more recent research findings in CMC (as well as neuroscience) sketched out by Ess. Again, the point here is that especially a Cartesian dualism that would radically separate our experience of the virtual from our experience of the real simply does not hold up—no matter the state of the technologies involved (i.e., whether the earlier text-based MUDs and MOOs or more recent environments as elaborated with 2D or 3D vision and sound). What we can now see perhaps more clearly: this is so because no matter where I go, so to speak, I always take my "I"—my unitary point of reference as an embodied being—with me.

Finally, Litska Strikwerda incorporates Sara Ruddick's phenomenological account of "complete sex," one that highlights the inextricable interconnection between the self as a unique person and identity and, *qua* embodied, its body and bodily actions (Ruddick, 1975, pp. 88f.). As Strikwerda points out, this issues in a decidedly anti-Cartesian account of bodies and sexuality as somehow radically divorced from the *persons*—in Kantian terms, the *autonomies*—somehow carried along (in Descartes' famous metaphor, like a pilot on a ship). This allows Ruddick, as Strikwerda makes clear, to thereby tie body and sexuality with the specific, unique *person* as an autonomous agent—and thus argue that "complete sex" can thereby reinforce the Kantian duty and virtue of always treating others as ends, never as a means only.

This specific linkage between phenomenology, Kant, and virtue ethics is both significant in its own right and as the occasion for turning to Kantian philosophy as the third major red thread that emerges here. To begin with, this linkage suggests a fundamental way in which we can reinforce a primary Kantian duty by way of phenomenology and virtue ethics. That is, Kant argues that a primary duty is to always treat the Other as an end, never as a means only—because this is demanded if we are to respect the Other as an autonomous agent, i.e., one capable of establishing its own ends, rather than being defined as the mere means to our ends. But where Kant developed this argument regarding a practical reason that was thereby distinct from a body—Ruddick's use of phenomenology to highlight the inextricable interconnections between the unique and distinctive *person* and that person's own body thereby extends the Kantian imperative to likewise always treat the *body* of the Other with the respect simultaneously required for the person *qua* moral autonomy.

At the same time, Ruddick's linkage between Kantian ethics and phenomenology thereby brings to the foreground the central importance of *virtue ethics* as a vital complement to more familiar Kantian deontology. Most briefly, virtue ethics highlights virtues as basic abilities or excellences that are requisite to both individual contentment and community harmony. These virtues are acquired only through long and sometimes difficult practice—as Shannon Vallor makes clear in her account of how we acquire the primary virtues of patience and perseverance, and thereby trust (2009, 9, cited in Ess, p. 9). These accounts emphasize that while we may recognize primary duties, beginning with respect for the Other (now understood as a single person inextricably intertwined with a specific body) always as an end from a purely rational standpoint using Kantian argument—our learning to act on these duties as embodied beings towards embodied beings thus requires long and often difficult *practice*. Specifically, we then further see this understanding of trust as a virtue in Carusi's and Myskja's accounts of trust as presumed and then enhanced in the virtuous circles they describe (cf. Ess, 2010). In this direction, we can further take up Grimen's action-oriented account of trust, as highlighted by Thorseth, as another form of such a virtuous circle.

At the same time, this linkage illuminates especially Myskja's account of *trust* as required specifically by the Categorical Imperative; that is, to treat the Other always as an end—because s/he is an autonomous being—thereby requires me to not only be truthful to that Other but also to *trust* that Other (Myskja, p. 123). In this light, our condition as embodied beings not only occasions the vulnerability and dependency that requires us to learn to trust one another in the first place (Løgstrup)—but insofar as embodied beings we are thereby inextricably interwoven with an "I" that incorporates (pun intended) a distinctive self as an autonomous agent (Ruddick). Kantian ethics thus entails two foundational moral duties: one, respect for the Other as an end, coupled with, two, the duty "...to act in accordance with the trust others place in us, that is, to act in a way that makes us deserving of trust" (Myskja, p. 124).

Trust, in short, is a given in the human condition and our intersubjective experience, a given practiced and enhanced through virtuous circles, especially in conjunction with the Kantian imperative to always respect the Other *qua* autonomy and *qua* embodied being as an end, never as a means only. When we go into virtual environments and virtual worlds, CMC research, Kantian philosophy, and phenomenology demonstrate that we always take our *selves*—our embodied identities as inextricably interwoven with the "I think"—with us. In this direction, both Kantian and phenomenological accounts, reinforced by CMC research over the past decade and as emphasizing identity as unitary, thereby illuminate Søraker's taxonomy in a particularly interesting way. That is, Søraker uses "indexicality" to refer to our sense of presence in a three-dimensional space, including a first-person view that, as we have seen, is rooted in Husserl. In his taxonomy, indexicality is definitive of virtual environments generally and virtual worlds specifically: only "the virtual" in the generic sense and virtual communities do not necessarily entail indexicality (see p. 85). This makes excellent sense for the sake of a philosophically coherent definition of these environments. But as human beings, we nonetheless appear to have an ineluctable tendency to bring a unitary sense of self with us, so to speak, wherever we go—including to the otherwise non-indexical domains of virtual communities *per se*. Again, as Ess summarizes, this tendency—with important and potentially liberatory exceptions, to be sure—is nonetheless now well-documented in CMC research on virtual communities. In parallel, the red thread of virtue ethics highlights: when we enter such environments and worlds, we should be mindful to practice our virtues as well, beginning with the virtues of respect for and trust in such Others.

More broadly, the Kantian thread runs through Carusi's arguments for trust in our shared, intersubjective environments (whether online or offline). Here we can further point out that Carusi draws on her earlier work on trust in virtual environments that draws on Kant's account of the aesthetic and the role of the *sensus communis* in his *Critique of Pure Judgment* (Carusi, 2008). For her part, Taddeo usefully invokes Kant's regulative ideal of a rational agent at the beginning of her account of e-trust. Finally, Thorseth makes use of Kant's notion of reflective judgment—also from the third *Critique*—to highlight the importance of "... taking account of the possible judgment of others," which thereby "requires the presence of, or potential communication with others" in her account of the communicative ideal to be achieved in deliberative democracy (p. 63).

In particular, Kant's reflective judgment intersects in important ways with the sort of deliberative judgment or *phronesis* that Socrates and Aristotle first highlight as a core facility for human beings as moral agents. *Phronesis* is marked by two key features. One, as Dreyfus emphasizes from his phenomenological perspective, *phronesis* is a kind of judgment that we can acquire and hone only within the embodied co-presence of others who exemplify this judgment, e.g., as experienced physicians, musicians, etc. (see Ess, this volume, p. 21). This is in part because *phronesis* requires elements of tacit knowledge that are known in and through the body (cf. Stuart, 2008).[4] Secondly, *phronesis* is not reducible to deductive and algorithmic processes, but rather comes into play precisely in those moments of (reflective) judgment when we must first choose between competing values and general principles that we can otherwise apply in a straightforwardly deductive fashion. In Kantian terms, *phronesis* is thus a key component and index of a radical autonomy—i.e., one that includes the freedom to make such judgments when we are caught between two (or more) different norms, and the difficulty is to discern which one(s) we should choose as most appropriate to our specific context.

Insofar as this is true, then, reflective judgment and *phronesis* are central points of contrast between autonomous human agents and artificial agents. Most simply, if *phronesis* cannot be implemented in artificial agents as computational devices dependent upon algorithmic processes—especially where such machines also lack the kind of tacit knowledge available to embodied knowers—then artificial agents, as lacking *phronesis*, would remain crucially distinct from human agents. This runs counter to Weckert's arguments in this volume that suggest that no significant differences exist between human and machine autonomy, so that both can be trusted in essentially the same sense. If, however, the *phronesis* affiliated with human autonomy remains beyond the capacities of computational devices, then we would have to distinguish more sharply the sorts of trust we owe to and hope for from humans vs. that which we could extend to machines.

While other significant and interesting connections and comparisons can (and will, we trust) be drawn between these chapters—we close with a summary outline of the larger philosophical account of trust and virtuality that emerges here. These chapters conjoin Kantian philosophy (not simply in terms of [communicative] reason, but also in terms of moral autonomy, aesthetic experience, reflective judgment, and the central ethical importance of respecting Others always as ends—which now includes *our* acting in trustworthy ways) with phenomenological analyses that emphasize our experience as embodied beings, as "body-subjects" (*LeibSubjekt*, Becker, 2001) who thereby experience our worlds (both virtual and non-virtual) from the unique perspective of "the I-think" as inextricably anchored in a distinctive (and mortal and vulnerable) body. These two philosophical perspectives—along with technological developments themselves—counter early Cartesian dualisms that would overly separate the virtual and the real. Given the robustness of trust—and its necessity for embodied beings dependent on one another—this blurring of the virtual and the real, at minimum, renders trust in online environments less problematic than, say, a decade ago. More positively, insofar as we can increasingly understand our experiences of the world—whether in virtual or non-virtual environments—as always experiences of a unitary and coherent (because, in part, embodied) self, we can see virtual environments as opening up new spaces into which we can productively extend our sense of being human as embodied, ethical beings for whom trust is initially a given. On these bases, virtual environments can thus counter early challenges to trust online—especially as such environments work to provide ways to "bootstrap" trust as our moral starting point, beginning with fostering the use of potential online deception in the direction of virtuous rather than vicious circles.

More broadly, this account reminds us to remember, so to speak, to bring our virtues with us—as we do our embodied identities—as we explore virtual worlds. Specifically, this account allows us to address the best and the worst possibilities of online environments. What we might think of as a Kantian-phenomenological philosophical anthropology conjoined with virtue ethics that emerges here thereby provides the ethical and philosophical resources to criticize entirely virtual child pornography (as well as more "mainstream" pornography) as nonetheless unethical as it works against a foundational equality norm. Positively, this philosophical anthropology helps us foreground the ways in which virtual environments may foster the best possibilities of online democracy and democratic deliberation—in part, precisely as these environments can allow us to foreground the legitimacy of the voices of women and children. Moreover, Strickwerda's critique of virtual child pornography and Carusi's and Myskja's accounts of the virtuous circles of trust highlight the importance of virtue ethics more broadly. These diverse implementations of virtue ethics suggest that as we increasingly interweave our lives with virtual environments, the virtues of embodied beings—including trust and its related virtues such as perseverance and patience—would seem to become all the more important for us to develop, rather than somehow less relevant.[5]

We suggested in our opening paragraphs that now is a very good time to revisit the issues of trust and virtuality, precisely because we now enjoy the advantage of considerable experience, empirical evidence, and, most of all, the time needed for careful, especially philosophical reflection. We hope this introduction has made clear how our contributors, both individually and collectively, flesh out this suggestion in multiple, rich, and specific ways. To be sure, our resulting accounts of trust and virtual worlds emerge as decidedly less revolutionary (because far less dualistic) than the 1980s' and 1990s' enthusiasms envisioned. For that, these accounts, as drawing especially on Kant and phenomenology as these further entail virtue ethics, thereby offer philosophical foundations that, we have argued, serve us better (where the "us" includes women and children, as well as men) as embodied beings than a 1990s Cartesian contempt for "meatspace" (where seeing the body as "meat" only simplifies the process of objectifying women and children, as well as men). These analyses further help us discern a number of important ways in which the initial challenges to trust of such disembodied environments can be overcome—thereby fostering a continued, if more qualified, enthusiasm for the potentials of virtual environments as now resting on philosophically more substantial grounds. In particular, the ethical and political upshots of these analyses—including a specific critique of virtual child pornography coupled with insight into how virtual environments may help bring to the fore the voices of women and children—stand as initial examples of the highly practical import and significance of these Kantian, phenomenological, and virtue ethics approaches.

None of this, however, is intended as a final and comprehensive account. In particular, as Søraker makes helpfully clear, some virtual environments will continue to provide ideal spaces for exploring more postmodernist and poststructuralist understandings of identity—understandings that retain crucial liberatory promise, not only for young people who are learning to negotiate the risks of online venues along the way towards developing their sense of self (cf. Livingstone, 2010, pp. 361ff.; Lüders, 2010)—but also for those whose sexual identities and preferences remain marginalized in real-world, "heteronormative" communities and societies (cf. Bromseth and Sundén, 2010). Along the way, however—insofar as these two philosophical foundations, as further highlighting the ethical imperative to *practice* primary duties to respect and trust Others as embodied and equal freedoms, thus engage our interest—we argue that these foundations thereby place us in much better positions to realize the multiple, positive promises and potentials of virtual worlds than especially the Cartesian dualisms that prevailed in earlier discourse and reflection.

We thus hope our readers will find this volume to inspire continued, if more critical but also well-informed, enthusiasm for and philosophical reflection upon virtual environments and their multiple potentials in contemporary and future realizations.

Acknowledgments

We wish to express our very great gratitude to the Programme for Applied Ethics, NTNU, for generous financial support—and to Marit Hovdal Moan for her conscientious and tireless labor in coordinating the many and complex logistical elements of the event. And, as several of our contributors note, all participants in the workshop deserve considerable thanks for their multiple contributions.

We are also very happy to convey deepest thanks to Mary Savigar and Steve Jones at Peter Lang—first, for their initial enthusiasm and encouragement for developing this volume, and then for their encouragement, advice, and constant support throughout its construction.

Notes

1. So Angel refers to the "anarchistic" exploits of the character Tom Collins, "...including the tale of his successful reprogramming of the M.I.T. virtual reality equipment to self-destruct as it broadcast the words, 'Actual reality, act up, fight AIDS'" (Larson, 2008, p. 79). As we will see, "actual reality" will become an important and technically precise term in Johnny Søraker's taxonomy of virtuality, developed in his chapter in this volume.
2. These concerns are not entirely theoretical, but are rather rooted in early and harsh experiences with what happens when such trust is broken. Ess points here (p. 12) to "Joan," an ostensibly disabled female participating in an early listserv devoted to women struggling with various disabilities. "Joan," however, was eventually uncovered to be "Alex," a male psychologist. Here we can note that the revelation evoked a strong sense of betrayal and loss of trust—including in the ostensibly utopian future that such online environments were thought to offer: "Many of us online like to believe that we're a utopian community of the future, and Alex's experiment proved to us that technology is no shield against deceit. We lost our innocence, if not our faith" (Van Gelder, 1985, p. 534, cited in Buchanan, 2010, p. 90). And as Buchanan further reminds us (2010, p. 88), in 1993, countering Rheingold's early enthusiasm for virtual communities, the now infamous "rape in cyberspace" also took place (Dibbell, 1993).
3. Here we can recall that Dreyfus' critique (2010) includes his drawing on the work of Kierkegaard, so as to emphasize the importance of *risking* judgments and practices in the

material world, where such risks have material consequences. This helpfully connects with the way in which risk is inextricably interwoven into our human condition more broadly and thereby the importance of *trust* as a kind of risk (so Løgstrup).

4. Indeed, recent comments by Stuart help clarify crucial connections between *phronesis*, phenomenological analyses of embodiment, and virtue ethics:

   > Aristotle thinks we need to embody the virtues. [...] We become virtuous by acting virtuously, that is, by acting in accordance with the virtues. This is the kind of reflective phronetic practice—when we choose between competing values and general principles—that hasn't yet been embodied, yet we do, in one sense, embody the virtues. They are embodied in living the *eudaimon*, the good life, about which we can speak and offer account through "told experience."

   Stuart goes on to emphasize how *phronesis*, *qua* embodied, thereby results in non-reflective habits and responses that nonetheless remain squarely rooted in reflective judgment:

   > However, there is another sense in which *phronesis* is embodied or incorporated; when we act with frequent regularity and the "right" action becomes, as we say, second nature. It has then been in-corporated, that is, brought "in" to the body as a characteristic of our lived experience. *Phronesis* is then no longer a matter of reflective judgement; at this stage has become realized as a matter of naturalised disposition.

   Stuart clarifies this last point via Husserl:

   > Let's think of Aristotle's virtue ethics in terms of Husserlian "I cans" (1989, pp. 270, 340, and others) and kinaesthetic skills resulting from dispositions and habits. Then we can see habituated social and, therefore, moral action, as phronetic dispositions which have become naturalized. We may begin by needing to reflect on how we should act, but our naturalised impulses become skilful non-reflective *phronesis*. So, for example, the skilful action, like spontaneously helping someone who has stumbles is an implicit phronetic "I can" requiring no reflection at all.

   (Personal communication to CE)

5. This point can be further supported and expanded upon by noting Miguel Sicart's use of virtue ethics in conjunction with computer-based games as significant instantiations of virtual environments and virtual worlds (2009).

References

Becker, B. (2001). The Disappearance of Materiality? In V. Lemecha and R. Stone (eds.), *The Multiple and the Mutable Subject* (pp. 58–77). Winnipeg: St. Norbert Arts Centre.

Borgmann, A. (1999). *Holding on to Reality: the Nature of Information at the Turn of the Millenium.* Chicago: University of Chicago Press.

Bromseth, J. and Sundén. J. (2010). Queering Internet Studies: Intersections of Gender and Sexuality. In M. Consalvo and C. Ess, editors, *The Blackwell Handbook of Internet Studies* (pp. 270–299). Oxford: Wiley-Blackwell.

Buchanan, E. (2010). Internet Research Ethics: Past, Present, Future. In Mia Consalvo and Charles Ess, editors, *The Blackwell Handbook of Internet Studies* (pp. 83–108). Oxford: Wiley-Blackwell.

Carusi. A. (2008). Scientific Visualisations and Aesthetic Grounds for Trust. *Ethics and Information Technology, 10*: 243–254.

Dibbell, J. (1993). A Rape in Cyberspace or How an Evil Clown, a Haitian Trickster Spirit, Two Wizards, and a Cast of Dozens Turned a Database into a Society. *The Village Voice*, December 21, 36–42.

Dreyfus, H. (2001). *On the Internet.* London and New York: Routledge.

Ess, C. (2010). Trust and New Communication Technologies: Vicious Circles, Virtuous Circles, Possible Futures. *Knowledge, Technology, and Policy*, 23 (3–4): 287–305.

Gibson, W. (1984). *Neuromancer.* New York: Ace Books.

Grimen, H. (2010). Tillit som senket guard [*Trust as lowered guard*]. In R. Slagstad (ed.), *Elster og sirenene* [*Elster and the Sirens*] (pp. 188–200). Oslo: Pax.

Heim, M. (1993). *The Metaphysics of Virtual Reality*. Oxford: Oxford University Press.

Husserl, E. (1989). *Ideas Pertaining to a Pure Phenomenology and to a Phenomenological Philosophy—Second Book: Studies in the Phenomenology of Constitution*. R. Rojcewicz and A. Schuwer, translators. Dordrecht: Kluwer.

Larson, J. (2008). *Rent: The Complete Book and Lyrics of the Broadway Musical.* New York: Applause Theatre and Cinema Books.

Pierre Lévy. 1998. *Becoming Virtual.* New York. Basic Books.

Livingstone, S. (2010). Internet, Children, and Youth. In M. Consalvo and C. Ess (eds.), *The Blackwell Handbook of Internet Studies* (pp. 348–368). Oxford: Wiley-Blackwell.

Lüders, M. (2010). Why and How Online Sociability Became Part and Parcel of Teenage Life. In M. Consalvo and C. Ess (eds.), *The Blackwell Handbook of Internet Studies* (pp. 456–473). Oxford: Wiley-Blackwell.

Løgstrup. K.E. (1956). *Den Etiske Fordring* [The Ethical Demand]. Copenhagen: Gyldendal.

Myskja, B. (2008). The Categorical Imperative and the Ethics of Trust. *Ethics and Information Technology, 10*: 213–220.

Rheingold, H. (1993). *Virtual Community: Homesteading on the Electronic Frontier*. Reading, MA: Addison-Wesley.

Rheingold, H. (2000). *The Virtual Community: Homesteading on the Electronic Frontier*. Revised edition. Cambridge, MA: MIT Press.

Ruddick, S. (1975). Better Sex. In Robert Baker and Frederick Elliston (eds.), *Philosophy and Sex* (pp. 280–299). Amherst, NY: Prometheus Books, 1975.

Sicart, M. (2009). *The Ethics of Computer Games.* London/Cambridge: MIT Press.

Stuart, S. (2008). From Agency to Apperception: Through Kinaesthesia to Cognition and Creation. *Ethics and Information Technology 10* (4): 255–264.

Thorseth, M (2008). Reflective Judgment and Enlarged Thinking Online. *Ethics and Information Technology, 10*: 221–231.

Turkle, S. (1995). *Life on the Screen. Identity in the Age of the Internet.* New York: Simon & Schuster.

Van Gelder, L. ([1985] 1991). The Strange Case of the Electronic Lover. In C. Dunlop and R. Kling (eds.), *Computerization and Controversy* (pp. 364–375). San Diego: Academic Press, 1991. (Originally published in: *Ms. Magazine,* October 1985, pp. 94–124.)

Weckert, J. (2005). Trust in Cyberspace. In R. Cavalier (ed.), *The Impact of the Internet on Our Moral Lives* (pp. 95–117). Albany: State University of New York Press, 2005.

Vallor, S. (2009). Social Networking Technology and the Virtues. *Ethics and Information Technology*. DOI 10.1007/s10676-009-9202-1.

Wellman, B. (2010). Studying the Internet through the Ages. In M. Consalvo and C. Ess (eds.), *The Blackwell Handbook of Internet Studies* (pp. 17–23). Oxford: Wiley-Blackwell.

SECTION I

Historical and Conceptual Perspectives

CHAPTER ONE

Self, Community, and Ethics in Digital Mediatized Worlds

CHARLES ESS

Introduction

The purpose of this chapter is to examine how the real and the virtual were initially conceived and understood in the early instantiations of virtual worlds (primarily MUDs, MOOs, and virtual communities) and correlative research in computer-mediated communication (CMC). These emphasize a strong dichotomy between the virtual and the real, and a correlative Cartesian dualism emphasizing the separation between mind and body—a dualism that, along with liberatory potential, thereby creates a central problem for establishing and fostering *trust* in online environments, insofar as such trust appears to require embodied co-presence with one another. I then trace the transition from such dualisms to more contemporary perspectives in CMC, communication theory, and philosophy (specifically information ethics and phenomenology) that emphasize how the real and the virtual (and with them, the mind and the body) are inextricably interwoven: such (re)turns to the body thus appear to restore important possibilities of establishing trust in online environments, especially as these environments under "Web 2.0" make possible more immediate representations of body (including gender). This initial framework is intended to set the broad stage for the specific thematic of trust in both offline and online environments in the chapters that follow in this volume.

Somewhat more carefully: in the first section, I begin a review of the history of approaches to the distinction between "the virtual" and "the real," starting with popular examples and research in the domains of computer-mediated communication (CMC) in the 1980s and 1990s. Here I seek to show how far this distinction was mapped through a strong *dualism*—i.e., an insistence not simply that these two entities are different, but so strongly opposed to one another that we are forced to choose between them in an "either/or" logic. This dualism was identified early on as Cartesian: I also argue that the science fiction novel *Neuromancer* (Gibson, 1984), as apparently underlying much of early conceptualizations of "cyberspace" and "the virtual," in fact draws on Augustinian and thereby ancient Greek and Gnostic dualisms. Such dualisms, stated harshly, demonize the body, sexuality, and thereby women in favor of a "pure" mind in a disembodied cyberspace.

While there are important liberatory potentials in such a disembodied space—the absence of the body creates immediate difficulties for establishing *trust* in online environments, as I explore in the second section.

In the third section, I return to the historical perspective, in order to show how 1980s- and 1990s-style dualisms between the real and the virtual, the offline and the online, are increasingly replaced by recognitions of how these once radically separate domains appear to rather seamlessly interweave one another, at least in the developed world. I do this by first taking up CMC research, beginning as early as 1995, that shows how virtual communities in fact depend upon and are intimately interwoven with their participants' offline identities and activities. Indeed, the cumulative judgment of many leading CMC researchers by the end of the first decade of this century is that the virtual/real distinction is no longer relevant to CMC research. In the second part of this section I turn to communication theory and information ethics as perspectives on media, self, and privacy that reinforce these more recent findings in CMC literature. Specifically, Walter Ong's understanding of electric media as introducing a "secondary orality" that *supplements* rather than replaces the affordances and skills affiliated with literacy and print argues that whatever cyberspace and virtual worlds may offer as new environments, the affordances and skills affiliated with these will complement rather than eliminate the skills and abilities of our more traditional, "real-world" selves. Finally, the third part of this section turns to phenomenology, as a specific aspect of the philosophical tradition that has been taken up within CMC research to highlight important ways in which our sense of identity, as rooted in body and gender, is largely reproduced, more or less accurately, in online environments. Through reviewing important figures and insights in recent phenomenology as applied to various instantiations of the virtual and the online, I argue that we can use phenomenological understandings of being human as a philosophical anthropology that helps explain how we behave online and offline as documented in CMC research. This is first of all as these phenomenological approaches stress the core importance of the body and embodiment in our ways of knowing and navigating the world.

This allows us to return to the question of trust and virtual worlds in section four. Most briefly, here I suggest that the (re)turn to the body and embodiment as thereby dissolving the once hard divide between the virtual and the real thereby suspends at least the most serious challenge to establishing and fostering trust online—namely, the lack of the body and our embodied presence with one another. With this as a starting point and orientation, we can then turn to the further exploration of trust and the virtual in the following chapters.

I begin by reviewing how both popular discussions and scholarly research surrounding "the virtual" and "the real" were initially mapped onto the online and the offline.

Descartes' Ghost: 1990s' polarities of "offline/online" and "real/virtual"

As has now been well documented (e.g., Ess, 2004; Lüders, 2010; Bakardijeva, 2010; Jensen, 2007), the first few years of scholarly and popular attention to the Internet and its multiple communicative possibilities were dominated by the assumption that the divide between "the virtual" and "the real"—as roughly co-extensive with "the online" and "the offline"—was best understood in what philosophers would characterize as an ontological *dualism*. That is, this dualism expressed itself precisely in the insistence that the worlds facilitated through online communication—including MUDs (Multi-User Dungeons), MOOs (MUD, Object Oriented), and virtual communities as instantiations of "virtual worlds"[1] —were radically distinct from and, indeed, largely opposed to "real life," e.g., as in the well-known acronym, IRL (in real life) as a reference to what online participants did when *not* engaged with one another via computer-mediated communication.

This opposition manifested itself in a wide range of ways. On a more popular level, an iconic, 1995 advertisement issued by the U.S. telecom company MCI famously characterized online communication as creating a realm of minds only, in which the harsh realities of well-known *differences*, including those of race, gender, and age, were erased (see Nakamura, 2000, for discussion). These sorts of hopes and dreams—characterized by some as "techno-utopianism"—were expressed perhaps most dramatically in John Perry Barlow's (in)famous "Declaration of Independence in Cyberspace" (1996). Again, a sharp—and as we are about to see, clearly Cartesian dualism—between the virtual and the real, between pure mind and mere matter, is presupposed:

> Cyberspace consists of transactions, relationships, and thought itself, arrayed like a standing wave in the web of our communications. Ours is a world that is both everywhere and nowhere, *but it is not where bodies live.* [...]
>
> Your legal concepts of property, expression, identity, movement, and context do not apply to us. They are based on matter, and there is no matter here.
>
> *Our identities have no bodies*, so, unlike you, we cannot obtain order by physical coercion. (1996, emphasis added, CE)

Slightly less bombastically, but certainly with equal enthusiasm, Howard Rheingold's *The Virtual Community: Homesteading on the Electronic Frontier* (1993) chronicled the multiple experiences and promises of "virtual community"—i.e., those interactions with one another made possible via CMC between humans otherwise not necessarily co-located or co-present with one another. In part thanks to Rheingold's glowing account, virtual community and communities—as thereby distinct from community in real life—quickly emerged as a primary focus of both popular enthusiasm and scholarly inquiry.

Apparently underlying these accounts of minds liberated in cyberspace and virtual communities, as radically divorced from (and, in Barlow's case at least, opposed to) "real life" as connected to the material, the physical, and *bodies,* was the vision of what such a cyberspace might be like as first articulated in William Gibson's profoundly influential novel *Neuromancer* (1984). From a contemporary perspective, much of Gibson's account is remarkably prophetic. But in one key way, as we will eventually see, his novel turned out to be fundamentally mistaken—i.e., precisely in the way it incorporated not simply a mind-body dualism, but, indeed, a mind-body dualism that directly echoes the Christian theologian Augustine. To begin with, Case, the main "cowboy" or hacker in the novel, loses his once virtuoso ability to navigate a disembodied cyberspace. Gibson describes this loss in the directly theological language of Augustine and his doctrine of Original Sin:

> For Case, who'd lived for the bodiless exultation of cyberspace, it was the Fall. In the bars he'd frequented as a cowboy hotshot, the elite stance involved a certain relaxed contempt for the flesh. The body was meat. Case fell into the prison of his own flesh. (1984, 6)

The Fall, as Augustine (re-)interpreted the second Genesis creation story, was precisely a fall from a kind of pre-sexual childhood innocence by way of the original sin of human disobedience. Influenced by Greek and specifically Gnostic beliefs in the soul as a kind of pure spark that "fell" from a (disembodied) heaven into the material world—precisely as a prison, as Gibson writes—Augustine further argued that this Fall manifests itself first of all in (an especially male) inability to exercise rational control over sexuality and thereby the body (Ess, 1995).

While the highly technologically mediated and secular future Gibson describes may seem very far from a 4th-century Christian theologian—in fact, Gibson echoes Augustine on this point even more directly later in the novel. An Artificial Intelligence attempts to distract—more precisely, seduce—Case from his primary mission by constructing an alternative space inhabited by a former girlfriend. Just as the woman, as reinterpreted by Augustine, is the sexual temptress who thereby occasions the man's Fall, so Case's former girlfriend tempts him back to the body *qua* "meat":

> There was a strength that ran in her, something he'd known in Night City and held there, been held by it, held for a while away from time and death.... It belonged, he knew—he remembered—as she pulled him down, to the meat, the flesh the cowboys mocked. It was a vast thing, beyond knowing, a sea of information coded in spiral and pheromone, infinite intricacy that only the body, in its strong blind way, could ever read. (1984, 239)

In short, early 1990s' dualisms, including Barlow's contempt for "meatspace," can be traced via *Neuromancer* to an Augustinian theology that is in turn rooted in Greek and Gnostic dualisms that would have us condemn the world, starting with bodies and sexuality (and, at least from a heterosexual male perspective, women) for the sake of a pure (i.e., disembodied) mind in a disembodied heaven—now secularized as "cyberspace" (see Ess, 2004 for further discussion).

At least the Cartesian version of this dualism did not escape notice entirely. While not widely acknowledged at the time, feminist scholars early on noted and critiqued these dualisms precisely as Cartesian. So, for example, Allucquere Roseanne Stone pointed out in 1991 that "that virtual community originates in, and must return to, the physical.... Forgetting about the body is an old Cartesian trick" (1991, 113). Somewhat later, as Janne Bromseth and Jenny Sundén note, feminist philosopher Rosi Braidotti (1996) likewise warned against

> ...the risks for women in buying into the notion of cyberspace as merely a place for subversive identity performances, freed from the limitations of the physical body. According to Braidotti, rather than liberate women this would repeat the Cartesian fallacy of separating mind from body. The dream of getting rid of the body reflects the understanding of masculinity as abstraction, she argued, of men as physically disconnected and independent, remapped onto cyberspace discourses. (2010, 271)

As we are about to see, Braidotti's connection between "subversive identity performances" and the Cartesian dualism prevalent in the 1990s thereby highlights an important conceptual element of both scholarly and popular attention that will, with the more recent shifts away from such dualism, likewise shift in crucial ways.

Finally, we can note here that Katherine Hayles (1999) traces Cartesian dualism back to the assumptions defining Norbert Wiener's foundational work on cybernetics and artificial intelligence. For her part, Hayles' conception of the "posthuman" includes a clear rejection of such dualism—and this in ways that will become thematic for research in the first decade of the 21st century:

> Emergence replaces teleology; reflexive epistemology replaces objectivism; distributed cognition replaces autonomous will; *embodiment replaces a body seen as a support system for the mind*; and a dynamic partnership between humans and intelligent machines replaces the liberal humanist subject's manifest destiny to dominate and control nature (1999, 288, emphasis added, CE; cf. Ess, 2005, 36).

The key point here (for our purposes, at least) is Hayles' highlighting the turn towards *embodiment* as a thematic focus that overturns the Cartesian (if not Augustinian) dualism still shaping the frameworks underlying much of the popular and scholarly discourse of the 1990s.

Trust in Disembodied Cyberspace?

If we take seriously the vision of achieving an ultimate liberation in a disembodied cyberspace, one of the immediate and most central challenges to arise is the issue of *trust*. As I discuss more extensively elsewhere (Ess, 2010b), philosophers have analyzed trust along three major lines: rationalistic or calculative accounts, affective accounts, and phenomenological accounts. For our purposes, the phenomenological approach is most salient, as it analyzes and identifies our most important experiences of trust as rooted in our embodied co-presence with one another.

While we may ordinarily take trust for granted (at least, until it's violated), it takes little reflection to see that trust is fundamental to human existence as such, and specifically to our experiences of friendship and democratic deliberation. Bjørn Mysjka helpfully begins this discussion by pointing to the work of Knut Løgstrup, who emphasizes first of all our *vulnerability* as embodied creatures. Most simply, we are dependent upon others to help satisfy basic needs—and, of course, we are, *qua* embodied, mortal: others can wound and hurt us by choosing to not help meet our needs—or, at the extreme, by seeking to end our lives. In the face of this vulnerability, we have little choice or hope but to *trust*: Myskja writes, "Trust is on the receiving end of ethical behaviour in the sense that trusting someone involves an appeal that they take responsibility for our well-being—but without any guarantee that they actually will ..." (2008, 214).

On Løgstrup's analysis, moreover, it is in our embodied encounters with one another that we can learn to *trust* one another:

> The personal meeting is essential for the ethical demand issued by human vulnerability, since we tend to form particular pictures and expectations of people we do not know, and are wary of them. These judgments will normally break down in the presence of the other, and this proximity is essential for the eradication of these preconceptions (Myskja, 2008, 214, with reference to Løgstrup, 1956, 22f.)

Given this essential importance of embodied co-presence for the sake of developing and sustaining trust, a first problem for any disembodied engagement with one another in online environments—whether in the form of an email from an unknown person or more elaborate forms of early text-based listservs and virtual communities—is: how am I to move beyond an initial wariness of the other to a posture of trusting them with my vulnerability as a dependent being?

As we will see in the next section, the problem of establishing trust in disembodied environments becomes specifically crucial in the context of virtual communities as well as other sorts of engagements in which it is critical for us to be able to *trust* that an online representation of the other person (more or less) faithfully coheres with their offline identities as embodied and thereby gendered beings. We can add here, finally, that the problem of trust online is further crucial from the perspective of both phenomenology and virtue ethics, as analyzed in the work of Shannon Vallor (2009). Very briefly, Vallor makes clear how the basic virtues of *patience* and *perseverance* are vital for our experiences of friendship, significant communication—and thereby *trust* (2009, 9). Echoing Løgstrup and drawing on the (phenomenological) work of Hubert Dreyfus and Albert Borgmann (see section 3.c), Vallor emphasizes that these virtues are learned primarily in the embodied world of face-to-face communication, which

> ...often forces us to be patient even when we would rather "tune out" or "switch off," to use telling metaphors. Yet this is precisely what builds patience as a virtue, rather than a grim resignation to the absence of an escape route from the conversation. (2009, 9)

Moreover, it is in the context of embodied co-presence that we are forced to confront "the gaze of the morally significant other": this gaze, Vallor continues, "... holds me respectfully in place and solicits my ongoing patience, [and] is a critical element in my moral development; though I might for all that ignore it, it creates an important situational gradient in the virtuous direction" (2009, 10). Specifically, Vallor argues that the absence of the embodied other in disembodied communication hinders our capacity to learn and practice *empathy* (as well as other virtues equally critical to individual and community well-being):

> Empathy is a cause for particular concern here, given that, as Danah Boyd notes, online communication can eliminate "visceral reactions that might have seeped out in everyday communication" (Boyd, 2008, 129). The editing out of some visceral reactions, such as anger or disgust, might aid online communication by encouraging greater trust and openness. But what if empathy is built upon visceral responses of its own? The ability to feel with others seems to require for its maximal development the capacity to identify *bodily* with another's suffering, even suffering not caused by a physical injury or illness. (2009, 10f.)

In parallel with Susan Stuart's appeal to contemporary findings in neuroscience (3.c, below), Vallor notes how this recent research underlines the importance of bodily co-presence for developing these sorts of virtues and capacities (2009, 11)

But again, this co-presence, along with the moral gaze of the other, is absent in a disembodied cyberspace. Insofar as these phenomenological accounts of the critical role of embodied co-presence in our development of patience, perseverance, empathy, and especially trust are correct—then clearly, an otherwise potentially liberatory cyberspace, especially as seen as a complete replacement for a despised "meatspace," seems to threaten us with the impossibility of trust and community rather than their ultimate realizations.

Beyond Descartes: Identity, embodiment, and the disappearance of the virtual-real divide

In this light, it is salutary that the Cartesian dualism defining much of the early discourse and reflection on cyberspace eventually gave way to more non-dualistic understandings of the relationships between the real and the virtual, the offline and the online. To see how this is so, I first sketch out some of the important ways in which a fundamental shift from Cartesian (if not Augustinian) dualism emerge in the literatures of CMC, focusing specifically on the thematics of identity play, gender and embodiment, and virtual communities. I then summarize how the shift away from Cartesian dualism is further supported by an important theory in communication, specifically with attention to Walter Ong's notion of "secondary orality," as reinforced by recent work in information ethics and CMC regarding our changing senses of self and privacy. Finally, I turn to recent developments in both phenomenology and neuroscience that reinforce the emphasis on embodiment and a blurred boundary between the virtual and the real that we will first explore in the context of CMC research.

CMC research: 1985–2010

Braidotti's reference to "subversive identity performances" makes explicit the connection between this early Cartesian dualism, the celebration of virtual communities as radically divorced from "meatspace," and a correlative popular and scholarly focus on postmodernist and poststructuralist theories of identity. That is, in opposition to modernist Western conceptions of the self,[2] postmodernist and poststructuralist conceptions rather emphasized our sense of self as involving multiple, fragmented, perhaps ephemeral identities—or, in Stone's characterization, as unstable, multiple, and contradictory (1995, cited in Bromseth and Sundén, 2010, 277). The *locus classicus* for this perspective is the work of Sherry Turkle (1995), who saw in the experiences of MUD and MOO participants instantiations of such postmodernist senses of identity and the multiple ways in which these venues were designed precisely to facilitate such identity play.[3]

As Bromseth and Sundén point out, perhaps one of the most fundamental ways in which we can play with our identity is in terms of gender and sexuality—where such identity as play is conceptually reinforced through the widely used work of Erving Goffman, who develops an account of the self as a series of performances as defined by specific relationships and social contexts, largely scripted for us by a larger society ([1959] 1990; 1974, cited in Bromseth and Sundén, 2010, 278; cf. Reading, 2009, 97–100). In this direction, we can note at least two of the most important ways in which such identity play can serve salutary, indeed, liberatory ends. To begin with, as Bromseth and Sundén carefully document, the possibilities of such identity play online are powerfully liberatory for those whose sexual identities and preferences are otherwise marginalized, if not demonized, by what they call "heteronormativity," i.e., the presumption of heterosexuality as the norm and only acceptable sexual practice (cf. 2010, 281). Secondly, especially recent research on Social Networking Sites (SNS) highlights the importance of such identity play for adolescents and young adults, despite the now well-known risks on online disclosure, in their exploration and formation of a developing sense of self (e.g., Lüders, 2010, Livingstone, 2010).

At the same time, however, these notions of a self as a bundle of multiple, ephemeral selves, along with the Cartesian split between body and mind that underlies them, were challenged in several ways in CMC literature, beginning with a very early event—originally reported in 1985—concerning gender and sex online. As described by Lindsy Van Gelder, "Joan" was an ostensible woman who participated in a listserv for disabled females. "Joan" gradually acquired the trust and empathy of the listserv participants, who revealed to "her" a range of highly intimate and personal reflections, anecdotes, etc. "Joan" was finally unmasked, however, as a male psychologist who was attempting through this deception to see if he could feel what it was like to be a woman—i.e., precisely the sort of online exploration of gender and identity that can hold important liberatory promise and potential. This bit of identity play, however, had devastating consequences for the women who had *trusted* "Joan" with considerable portions of their personal lives (Van Gelder, 1985, 534, cited in Buchanan, 2010, 90). This makes the point that even in the earliest online environments, limited as they were to textual representations of self, an authentic representation of our offline, embodied gender remains crucially important for online communication, especially for the sake of developing *trust* and community.

Indeed, by 1998, CMC researchers Beth Kolko and Elizabeth Reid highlighted the importance of these fundamental expectations as a crucial *problem* for online communities. As paraphrased by Bromseth and Sundén, Kolko and Reid argue that gender is crucial as

> ...a central resource for how we make sense of reality; how we interpret and relate to other people. In physical contexts, the body is used as a central resource in this interpretative interactional work. To be culturally intelligible online, it is even more crucial to give a coherent gendered self-presentation in creating credible selves... (Bromseth and Sundén, 2010, 277)

Online identity play is thus somewhat Janus-faced. To be sure, it opens up important, potentially vital liberatory possibilities, especially for those otherwise marginalized in their offline communities. At the same time, however, insofar as our offline bodies, gender, and sexual preferences and practices may thus be easily disguised online—such potential for disguise creates central problems for trust and community.

Moreover, Nancy Baym's work on online communities demonstrated as early as 1995 that then prevailing presumptions of virtual communities as somehow radically divorced from their embodied constituents simply did not hold up. Rather, her analysis of fan communities made clear that our online behaviors were directly interwoven with our offline identities and practices. Soon thereafter—and directly contrary to the MCI advert's depiction of a bodiless cyberspace free of gender, race, and age—Susan Herring's discourse analyses of online communication demonstrated the predominance of gender-specific communication characteristics, even in the instances of authors trying to "pass" as differently-gendered (1996, 1999). By the same token, Beth Kolko and colleagues could make painfully clear by 2000 that race, like gender, remains stubbornly present in cyberspace (Kolko, Nakamura, and Rodman, 2000). In the face of this and similar research, it is perhaps not surprising that by the first years of this century, two of the most significant voices for the 1990s' enthusiasm for virtual worlds as radically divorced from our embodied lives revised their earlier accounts: Rheingold, in particular, warned against the dangers of such sheer virtuality for an engaged political life in our offline domains (2000; cf. Bolter, 2001).

Finally, as Marika Lüders and others have pointed out, Turkle's work, however influential, nonetheless focused on a rather limited sample of Internet users—i.e., precisely participants in role-playing games facilitated by MUDs and MOOs (2007, 9–12). But, as access to and ways of using the Internet rapidly expanded throughout the 1990s and the first decade of the 21st century, very clearly, who goes online for what purposes has likewise dramatically expanded beyond those who enjoy the role-playing possibilities of MUDs and MOOs. At least in the developed world, more and more of us are "networked individuals" (Wellman and Haythornthwaite, 2002) who know how to use the Internet and its various applications to do everything from communicate with friends and colleagues via email and social networking sites to checking bank balances and shopping online. By 2005, in fact, Maria Bakardjieva could write of Internet society as "The Internet in Everyday Life" (cf. Bakardjieva, 2010).

Such an everyday internet means specifically less emphasis on the once prevailing poststructuralist and postmodernist understandings of identity as multiple, fragmented, and ephemeral and greater emphasis on identity as defined by an embodied (and thereby gendered) being. So Lori Kendall found by 2002, echoing Herring's work, that participants in the *BlueSky* community of her study

> ...brought their offline understandings and expectations about gender to their online interactions. As in people's offline relationships and communities, *BlueSky* participants enacted and constructed gender identities through their online interactions, asserting gendered identities, and, in some cases, arguing about what gender means. (2010, 320).

More broadly, Kendall finds that "...in most long-standing communities, [identity] deception is minimized. The formation of community depends upon consistent identities" (2010, 319). In particular,

> ...most people in virtual communities wish to represent themselves in consistent and realistic ways. People do manage to perform consistent identities online. Among other things, this means that the aspects of identity that some hoped would become insignificant online—such as race, class, and gender—remain salient. (2010, 319)

Kendall explains this pattern in part by way of reference to the work of Albert Borgmann, who writes that "...in the end and deep down...we crave recognition, the acknowledgement of who we are in fact" (2004, 59, cited in Kendall, 2010, 319). As we have already started to see (2) and will explore more fully below (3.C), Borgmann's work is part of a larger series of phenomenological analyses that emphasize precisely the role of the body and embodiment in grounding our sense of identity. More broadly, *contra* the Cartesian fantasy as depicted in the MCI advert of a bodiless cyberspace devoid of the material and the bodily—for better and for worse, our online lives appear to be increasingly interwoven with our offline identities, meaning first of all, our identities as embodied and thereby gendered beings.[4]

What this means, finally, is that more contemporary CMC research on community now brings to the foreground multiple ways in which virtual communities are tightly interwoven with their offline constituents—so much so that the character and focus of contemporary CMC research has likewise changed:

> Most communities connected through the Internet involve both online and offline components. Even in virtual communities that primarily exist online, participants often seek to meet one another face-to-face. Meanwhile, many offline groups seek to enhance their communities through online participation. In recent research on community and the Internet, the emphasis is shifting from ethnographic studies of virtual communities, to studies of people's blending of offline and online contacts. (Kendall, 2010, 320)

Communication theory and information ethics

These important shifts in CMC research resonate with a significant theory in communication, especially as amplified through insights regarding changing notions of privacy as explored in information ethics.

The communication theory emerged over the latter half of the 20th century, beginning with the work of Harold Innis, Elizabeth Eisenstein, Marshall McLuhan, and Walter Ong, and was subsequently enhanced through the work of Naomi Baron (2008) and Zsuzsanna Kondor (2009). Broadly, this theory discerns important correlations between four distinct "communication technologies," beginning with *orality*, and conceptions of self and community. *Orality*, as affiliated with preliterate and pre-agricultural societies, relies on repetition, rhyme, and performance (e.g., music and dance) as ways of preserving the spoken that would otherwise evaporate into the air. As is well-known, some of the most important cultural accomplishments in both Western and Eastern societies were initially composed and preserved as oral performances—e.g., *The Iliad* and *The Odyssey*, the various community recollections that eventually emerged as Scripture for both Jews and Christians, the *Qu'ran,* and the *Analects* of Confucius, just to name a few. *Literacy* first emerges alongside the rise of agriculture, with writing initially developed for the sake of accounting granary contributions, debts and payments, etc. in the first cities. Once we can "freeze" the oral into a more permanent form, this facilitates the first critical reflection we know as philosophy among the ancient Greeks—and with it, the development of formalized notions of logic, as first made articulate in the writings of Aristotle (cf. Baron, 2008, 196f.) The invention and rapid spread of *print*, as thereby helping to diffuse *literacy* among ever-growing numbers of people, is affiliated with, for example, the Protestant Reformation: only when the Bible becomes widely available—and in a fixed, standardized edition (in contrast with the 3,000+ variant manuscripts to be found in medieval libraries)—does it make sense for a Martin Luther to insist that our way to God is *sola scriptura*, [through] Scripture only. That is, whereas written texts were earlier constructed as something like a symphony score, i.e., to be performed, heard, and interpreted in *community* (cf. Baron, 2008, 185)—*sola scriptura* emphasizes the importance of the *individual* reading and interpreting Scripture as a primarily solitary activity. Similarly, modern philosophical and political concepts of the individual—understood specifically as a reflective, *autonomous* entity capable of radically free choice whose *right* to choice both justifies and requires the modern liberal-democratic state—seem to closely depend upon the *affordances* of literacy-print as communication modalities (see Chesebro and Bertelsen, 1996; Baron, 2008, 196f.).

In this direction, we can note Foucault's late work (1988) on reflective writing—e.g., diaries and letters—as first emerging among the Roman elite of the 1st century as "technologies of the self," i.e., as communication modalities that foster the *individual* construction of the self as a primary *virtue* or excellence of the "soul" (cf. Lüders, 2007, 48). Foucault's analysis extends from Socrates and the Stoics through Freudian analysis, and thereby extends the understanding first developed by Innis et al. with regard to how literacy and then print help foster an especially modern sense of the individual (Bakardjieva and Gaden, 2010). As a last example along these lines, Baron describes the rise of the Commonplace Book, beginning with the Renaissance humanist Erasmus of Rotterdam. Young gentlemen were encouraged to collect in such a book the best and most striking examples of *writing* that they encountered, with a view of then incorporating these best thoughts into one's own thinking and writing—i.e., as an *individual* project facilitated by the skills and affordances of literacy and print. At least through the end of the 19th century, keeping such a Commonplace Book was, to borrow Foucault's phrase, yet another technology of the self that depended on the skills and affordances of literacy and print—helping to produce along the way a distinctively modern sense of the self as an autonomous individual (cf. Baron, 2008, 196f.).

All of this changes yet once again with the rise of what McLuhan and Ong characterize as *electric* media, beginning with radio, movies and then TV, which in turn introduce a new set of media-related sensibilities, affordances, and skills. Ong characterized these in terms of what he called "secondary orality," insofar as an electric medium allows us

> ...to reintroduce the immediacy of oral communication, brings sound and gesture back into the human sensorium, and changes written text from something that is fixed and unchangeable to something malleable, or as Richard Lanham puts it, "volatile and interactive" (1993, 73, in O'Leary & Brasher, 1996).

Such transformations are especially manifest in the era of "Web 2.0," i.e., as we are increasingly able to produce and distribute media content of our own making through the Internet, e.g., in the form of YouTube videos, online video-conferencing, etc. Zsuzsanna Kondor has helpfully characterized these media possibilities as thereby involving a shift to "secondary literacy." This term highlights the important fact that texts and literacy remain significant components of our various ways of communicating through new media (e.g., SMS, chat, blogs and micro-blogging such as the Facebook update and Twitter, etc.). From Kondor's perspective, this further means that the *rationality* affiliated with literacy will not necessary be lost, but rather,

> Now there is no need to constantly translate or encode experiences and ideas into verbal, and thus propositional, structures. This possibility increasingly opens the floor to the idea of perceptual and motor processes which do not need permanent conceptual supplementation (though sometimes conceptual apparatuses might facilitate responses). (2009, p. 180)

Specifically, for Kondor this development represents a salutary *enhancement* of the sense of rationality affiliated with literacy-print, insofar as our ability to communicate and represent ourselves in aural and visual ways, not simply textual, thereby helps bring back into central focus the *body* as our primary locus and expressive agent of our sense of self (2009). In this way, as we will see below, Kondor's account foregrounds recent media developments with the (re)turn to *embodiment*—a (re)turn that is crucial for more contemporary understandings of the correlations between new media and our sense of selfhood and identity.

Completing the logic of this framework—the shifts we now see underway towards greater emphasis on the communication technologies and modalities of secondary orality-literacy, the theory predicts, will bring in their train correlative shifts in our sense of self. Very briefly, especially if we stress the *orality* side of secondary orality, we would hence expect a (re)turn to the more *relational* sense of self affiliated initially with oral cultures. Such a self—as manifest in ancient Confucian thought, as well as in traditional Japanese, Thai, African, and other indigenous cultures—understands itself primarily in terms of the *relationships* that define it. This is to say, on this view, I *am* my relationships—e.g., as spouse, friend, father, child, brother, uncle, teacher, scholar, member of larger communities, etc. (Ames and Rosemont, 1999; Hongladarom and Ess, 2007). In ways that anticipate and are thus reinforced by Goffman's notion of the self as performative, the relational self is thus understood to indeed show a different "face," depending precisely on the primary relationship(s) that are in play in a given (and socially defined) context (Goffman, [1959] 1990; see Lüders, 2010, 457).

As I have argued elsewhere, the very great weight of available research in the fields of CMC and information ethics makes very clear that we are indeed witnessing such a shift in our sense of self (Ess, 2009, 2010a). One of the most significant markers of this shift is in changing expectations and practices regarding individual privacy, including data privacy. This can be seen in a range of ways, beginning with what Anders Albrechtslund (2008) has characterized as our "participatory" or "voluntary" *lateral* surveillance, in contrast with the more familiar "top-down" surveillance of the Orwellian Big Brother kind. By these terms, Albrechtslund denotes our willingness to share in various online fora such as social networking sites (SNS) and YouTube information that, until a few years ago at least, was considered a matter of *individual* privacy, i.e., not to be shared in a public, or even semi-public form.

A striking example of such willingness to share is the "pocketfilm" (i.e., as shot with a mobile phone), *Porte de Choisy*, the first-place winner of the 2007 Pocketfilm Festival (see <http://www.festivalpocketfilms.fr/article.php3?id_article=648>). The film, taken by a boyfriend of his girlfriend in their apartment, begins with the young woman talking on her mobile phone while apparently sitting on the toilet. The episode includes an apparent sexual encounter, followed by the young woman sitting naked at her computer, attempting to finalize directions for meeting someone at the Porte de Choisy.

As Gabriella David points out in her analysis of the film, modern conceptions of *individual* privacy—including privacy in the bedroom and the bathroom—simply did not exist in the Middle Ages (2009, 79). Rather—consistent with the

Innis-Eisenstein-McLuhan-Ong-Baron-Kondor account—such individual privacy for the sake of sexual intimacy and bodily needs emerged alongside modern notions of the individual, culminating in the 19th and 20th century notions of the bedroom and the bathroom as private spaces *par excellence*. The film *Porte de Choisy*, however, "violates" both of these spaces simultaneously. To be sure, so-called reality shows such as *Big Brother* would suggest that at least some portion of our appetite for such self-disclosures is driven by simple voyeurism, the tantalizing pleasures of seeing exposed what is otherwise hidden and concealed, etc. And seen from the perspective of more traditional understandings of the self and privacy, the shooting and then publication of *Porte de Choisy* on the internet are hard to see as anything other than what Kurt Röttgers calls "the pornographic turn," in which "...the educational process of making a soul has become a never-ending process of the post-modern individual and has been transferred to these systems of surveillance" (2009, 89). Röttgers further argues that a "total transparency" amounts to the total loss of decency, which includes for him a necessary politeness towards and respect for other people (2009, 90).

Perhaps so—but only as long as we retain modernist notions of the individual and then of individual privacy. *Porte de Choisy*, however, dramatically illustrates that these notions of privacy no longer hold—that we have entered, as David points out, the era of what McLuhan presciently identified in terms of *publicy*, i.e., a sense of privacy that is no longer defined by a simple individual-private/social-public dichotomy, but exists somewhere between these two poles. In the case of *Porte de Choisy,* David sees "publicy" as instantiated by what she characterizes as the "exteriorized intimacy" in the film as made and then made publicly available (2009, 86). Such a shift in privacy makes sense, I have argued, if our understanding of ourselves is likewise changing from that of the modernist individual to a more "post-post-modern" relational self (Ess, 2009, 2010a).

Along these latter lines, Hille Koskela (2006) argues that as *we ourselves* produce such visual images and distribute them publicly across the internet, we thereby claim an "...active agency—a condition in which [we] can be subjects rather than objects (of surveillance)" (2006, 172, cited in Reading, 2009, 96). Specifically, for our purposes, Koskela characterizes this sort of self-exposure as a political act—one which, as characterized by Reading, "involves rethinking conventional binaries of object/subject/virtual/embodied, as well as the distinction between material and virtual space" (*ibid*, 97). That is to say: *Porte de Choisy* exemplifies a blurring of the once dichotomous boundary between virtual and real, the online and the offline.

Perhaps somewhat less dramatically, it now seems clear that especially young peoples' uses of SNSs such as Facebook involve a sense of privacy that lies somewhere between purely individual privacy, on the one hand, and the public on the other. That is, young people routinely reveal aspects of themselves—including their sexual interests and activities—among at least a small circle of friends, being careful (so far as possible) to ensure that this sort of information is *not* more widely available (e.g., to mom and dad: see Lüders 2010 for examples and discussion). Again, whatever their elders may think of all of this—it seems clear that this new sense of "friend-privacy" represents a shift from a 20th-century individual sense of privacy to a much more *relational* self for whom "privacy" refers to the privacy of a close circle of friends.[5]

In light of these now manifest changes, a key question now becomes: will secondary orality-literacy (and with it, the relational self that they foster) *supplement*—or, more radically, *replace*—the skills, affordances, and abilities fostered through literacy-print (and with them, the individual self affiliated with modern liberal-democratic societies)?[6] The force of this question becomes even greater if we further see it in the light of a specific connection between the Innis-Ong schema and 1990s' enthusiasms for the liberation of disembodied minds in a virtual cyberspace radically divorced from "real-life." Very simply, at least some of the early enthusiasm for "the virtual," as in the form of virtual communities and "cyberspace" more broadly was driven by an appropriation of the Innis-Ong schema that first of all connected (more or less correctly) the virtual and cyberspace with the technologies and affordances of Ong's secondary orality. So, for example, Stephen D. O'Leary and Brenda Brasher observed that electronic media, precisely as affiliated with Ong's secondary orality, would no doubt entail an enormous cultural shift: "The transformation to secondary orality is no less momentous than the shift from primary orality to literacy, and the full implications of this transformation will take centuries to appreciate" (O'Leary & Brasher, 1996, 256). In the then highly influential view of pundits such as Nicholas Negroponte, this shift would be *revolutionary*: electronic media would render print, specifically books, obsolete (1995).[7]

From the standpoint of the Innis-Ong schema—Negroponte (and, for that matter, John Perry Barlow)—would be correct *if* secondary orality as a modality of communication were to *replace* literacy-print. For his part, however, Ong rather envisions that secondary orality will follow the pattern of the previous two stages of literacy and then print: that is, each of these emerged ultimately as a *supplement* and enhancement of the previous stage, not its abolition. If Ong, rather than Negroponte and Barlow, is correct, we would then anticipate that secondary orality-literacy will supplement rather than render obsolete the modalities of literacy and print—and with them, the sort of individual self they are affiliated with in modernity. By the same token, we would expect that "the virtual" and "the real," the online (as predominately the communicative space of secondary orality-literacy) and the offline (as including the spaces and practices devoted to literacy-print) would likewise settle into a relationship of complementarity, rather than standing as hard, oppositional binaries forcing us to choose the one at the cost of the other.

I have argued that there is good evidence that Ong is correct in these ways: first of all, insofar as we can observe in both Western and Eastern contexts the emergence of a kind of hybrid self, one that conjoins the characteristics, practices, and communicative skills affiliated with both an individual self (and thereby literacy-print) and a relational self (and thereby secondary orality-literacy: see Ess, 2009, 2010a). As well, Ong's understanding of the complementary relationship between secondary orality and literacy-print is consistent with the shifts we have seen sketched out above from a 1990s dualism emphasizing a hard dualism between the virtual and the real to more contemporary views that highlight instead precisely the ways in which the offline and the online cohere with one another—mediated, so to speak, through an embodied and gendered person who remains at the center of his or her agentic and performative self.

In sum, both from the historical perspective of research in the domains of CMC and the perspectives developed through the communication theory that emerges across the work of Innis through Kondor, coupled with recent research on privacy both within information ethics and, again, CMC—it appears that the 1990s' tendency to map the distinction between the virtual and the real, the online and the offline, in strongly (Cartesian) dualistic ways was, at best, a temporary view. As captured, for example, in McLuhan's neologism *publicy* (i.e., as a cross between public and privacy) and Luciano Floridi's term "onlife" (2007), developments over the first decade or so of the 21st century rather point to how the offline and the online—and with them, at least much of our experience of "the virtual" and "the real"—are increasingly seamlessly interwoven with another, rather than dichotomously opposed.

Philosophical trends and developments

We have seen that among the first to object to the (re)appearance of Cartesian (if not Augustinian) dualism in 1980s' and 1990s' conceptualizations of cyberspace and the virtual were feminist scholars such as Allucquere Stone and Rosi Braidotti. Stone, in particular, warned against "forgetting the body" as an "old Cartesian trick (1991, 113). Similarly, Katherine Hayles explicitly enjoined *embodiment* as a counter to the Cartesian dualism she discerned in the foundations of cybernetics and at least early work on Artificial Intelligence. We further saw, *contra* 1990s' hopes for a disembodied cyberspace in which gender, race, and age were no longer visible, body and gender as core components of identity become crucially important in the constitution of virtual communities, especially as these are analyzed from the more recent perspectives that emphasize the interdependence between the online and the offline ("onlife"). In particular, Lori Kendall invokes Albert Borgmann's insight—rooted in Borgmann's larger phenomenological perspective—that "...we crave recognition, the acknowledgement of who we are in fact" (2004, 59, cited in Kendall, 2010, 319). Finally, this intersection between feminist approaches to CMC research on virtual communities, as stressing the importance of body and embodiment and phenomenology is further in play in the work of Zsuzsanna Kondor and her analysis of contemporary communication media as highlighting a secondary literacy that complements rather than replaces earlier forms of literacy and rationality.

These intersections between CMC research, communication theory, and phenomenology are salutary but not accidental. That is, if we turn briefly to more directly *philosophical* approaches to contemporary media technologies, we can notice an important range of philosophers working out of the phenomenological tradition who, precisely through their focus on embodiment, offer a wide range of observations and insights that are specifically relevant to our concern with the virtual, the real, and their possible interrelations.

To begin with, Albert Borgmann's influential philosophy of technology is squarely rooted in a phenomenological insistence on the centrality of the body: "The human body with all its heaviness and frailty marks the origin of the coordinate space we inhabit. Just as in taking the measure of the universe this original point of our existence is unsurpassable, so in venturing beyond reality the standpoint of our body remains the inescapable pivot" (1999, 190). Such a body is marked specifically by *eros* and *thanatos*—we are, indeed, mortal, whatever dreams of cybernetic immortality may motivate proponents of a putative liberation in a bodiless cyberspace. Echoing the work of Løgstrup and anticipating what will become a significant theme in contemporary CMC research, Borgmann specifically warns against the elimination of the body in cyberspace as leading all too easily to the reduction of *persons* to commodities (1999, 200). Finally, Borgmann uses his phenomenological account of a bodily centered way of coming to know and learning how to navigate the world—including the social world—as the basis of a critique of distance or online education.[8]

Hubert Dreyfus first took up phenomenology as a way of developing his highly influential critiques of especially "hard" Artificial Intelligence predominating in the 1960s through the 1980s (1972, 1992). In parallel with Borgmann's critique of distance education as disembodied education, Dreyfus extended his earlier analyses to argue that the most important sorts of education—specifically, education that seeks to foster the unique capacities of *phronesis* as a distinctive kind of *judgment*—cannot be accomplished in disembodied contexts (2001). Rather, as our practices of apprenticeship—e.g., as musicians in a small class with an acknowledged master of our instrument and style, as advanced students working closely with a senior teacher, or as medical students on rounds with experienced physicians—exemplify, we learn not only specific sorts of knowledge, techniques, and skills, but, most importantly, we begin to develop the sort of judgment needed to apply those abilities appropriately and effectively within a given (and always distinctive) context.[9] At the same time, Dreyfus draws on Kierkegaard to argue that virtual domains, especially as they eliminate both the vulnerability of our bodies and thereby the risks that we take when we encounter others in real-world, face-to-face contexts, thereby *prevent* us from learning how to "come to grips with" (note the reference to body in the phrase) with vulnerability, risk-taking, and thus *trust* as central components of our being human with one another (2001). In this way, we can note, Dreyfus' use of Kierkegaard thereby echoes Løgstrup's emphasis on our embodied encounters with one another as experiences of vulnerability through which we eventually figure out how and when we might trust one another.

Similar and important critiques of a disembodied cyberspace—critiques that begin by bringing to play rich phenomenological accounts of how we know and experience the world as embodied beings—are likewise developed by the philosophers Barbara Becker (2001) and Darin Barney (2004). More recently, Susan Stuart has conjoined such phenomenological foundations with contemporary findings in neurosciences—specifically, an understanding of how we know and engage the world through our bodies called enactivism. In contrast with more Cartesian views of a sharp mind-body split at work not only in 1990s' perspectives on cyberspace and the virtual, but also in hard AI and "cognitivism" (the view that the mind is essentially a computer that manipulates symbolic representations of various components of the world around us), enactivism foregrounds various *pre*reflective and *non*cognitive ways in which our bodies, in effect, constantly interrogate and discern how to respond to our immediate environments. Once again, *contra* Cartesian dualism, Stuart's account shows that "there is an inseparability of mind and world, and it is *embodied practice* rather than *cognitive deliberation* that marks the agent's *engagement* with its world" (2008, 256).

Somewhat ironically, perhaps, as enactivism thus emphasizes the noncognitive and nonrepresentational dimensions of our knowing the world through our bodies, it thereby highlights as a positive what for Gibson was a negative, i.e., the body as "...a vast thing, beyond knowing, a sea of information coded in spiral and pheromone, infinite intricacy that only the body, in its strong blind way, could ever read" (1984, 239).

For our purposes, these phenomenologically based analyses of how we know and navigate our world—including our social worlds and the communicative possibilities they entail, from orality through secondary orality-literacy—thus provide a strong philosophical framework for approaching the possible relationships between the online and the offline, the virtual and the real. These approaches, as emphasizing the body and embodiment as core to our sense of self, our ability to discern and move appropriately in material/social domains, directly counter especially postmodernist and poststructuralist accounts of the self as multiple, fragmented, ephemeral, etc. More broadly, these approaches decisively (in my view) put to rest the remnants of Cartesian (and, I would hope, Augustinian) dualisms that still pervaded much of our perspectives on cyberspace and the virtual/real distinction in the 1990s. In so doing, these analyses thus reinforce and expand the findings of CMC research on the virtual as exemplified in virtual communities in the first decade of our century: that is, they provide what we might think of as the philosophical anthropology that is consistent with the findings in CMC research that return to the center of our attention the crucial importance of authentic representations of body and gender in our online engagements with one another. By highlighting our identity as inextricably interwoven with our bodies—as mortal and gendered, and thereby as distinctive and unique—these approaches further provide philosophical resources, i.e., a philosophical anthropology, including an analysis of a lifeworld that is not reducible to markets and commodity exchanges, that might help us resist the otherwise overwhelming pressures to commodify ourselves as exchangeable goods online, e.g., as we work to "brand" our identity through a SNS profile or blog.[10] Most importantly, for our purposes at least, these approaches offer strong reasons *why* the Cartesian split between the virtual and real has collapsed—or, stated differently, why the material and the bodily have apparently resisted the once urgent calls for liberation via escape into a bodiless cyberspace. Simply, as foundational to our sense of self and our engagement with others and our world, the body (and with it, the material) remains, in Borgmann's phrase, "the origin of the coordinate space we inhabit" (1999, 190).

To say this slightly differently: the collapse of the Cartesian distinction between the virtual and the real is not simply because, as Klaus Bruhn Jensen has perceptively put it: "Old media rarely die, and humans remain the reference point and prototype for technologically mediated communication" (2010, 44). More fundamentally: from such phenomenological perspectives, "We are essentially bodily creatures that have evolved over many hundreds of thousands of years to be mindful of the world not just through our intellect or our senses but through our very muscles and bones" (Borgmann 1999, 220).

From Cartesian minds in cyberspace to embodied participants/constructors of "on/life": The dissolution of the virtual/real boundary and implications for trust

In short, over the past decade (if not a little longer), CMC research has extensively and intensively documented the multiple ways in which the offline and the online more and more seamlessly interweave with one another rather than stand in sharp, 1990s-style opposition. This does not necessarily mean that "the real" and "the virtual" are no longer indistinguishable—though, as Susan Stuart observes, the day is coming in which it will be increasingly difficult for us to tell the difference between our experiences in "the real world" and what will be increasingly more sophisticated "virtual worlds," including, perhaps, the worlds made possible by her "virtual reality adultery suit" (2008).

It does mean, however, that the 1990s' presumptions of a hard boundary between the virtual and the real simply no longer hold—and thereby, many of the initial claims and worries made about life online more or less collapse, for better and for worse. Most broadly, any putative "liberation in cyberspace," as an escape from "meatspace" (Gibson, 1984; Barlow, 1996) seems to be largely "techno-utopian" fantasy. We can deeply regret that the failure of some sort of "pure" cyberspace—at least one in which race, age, and gender ostensibly disappear—to materialize means that, among other things, women will remain likely victims of traditional gender stereotyping, of the male gaze, and, worst-case, of the male violence against women that is legitimated by such traditions. At the same time, however, let us recall the Augustinian roots of the dualism and contempt for the body—if not for the world (*contemptus mundi*)—that philosophically underlies the 1990s' hopes for liberation in cyberspace. This Augustinian contempt has legitimated unfathomable damage across the millennia, not simply in its role in justifying violence against women, but also in justifying damaging exploitation of the natural order more broadly (Ess, 1995). In this light, the research findings in CMC and related fields (including, as Susan Stuart and others point out, our increasing understanding of the brain and body) represent a salutary shift away from such contempt.

More narrowly, these more recent understandings of how the online and the offline increasingly interweave themselves in our time are of considerable significance for the central thematic of *trust*. As we have seen, there are strong threads in the philosophical literature on trust that highlight the role of the body and embodiment in our learning to trust one another. Again, as Bjørn Myskja has pointed out, the Danish philosopher and theologian Løgstrup articulates how we learn to overcome our *distrust* of one another by "reading" (my term) the Other as an embodied being, so to speak, in front of us. In light of the way in which *disembodiment* online is a primary obstacle to our establishing and fostering trust with one another in online venues and environments, the return of the body—e.g., as more directly re-presented via video and audio in the various venues and modalities made possible by Web 2.0—thereby reduces online disembodiment and at least increases the possibilities of our re-presenting our bodies (including facial gestures and other components of nonverbal communication) in ways that may help establish and foster trust.

To see how this is so in much greater detail and from a wide range of perspectives is now the work of the following chapters.

Notes

1. To be sure, there are important additional instantiations of "the virtual" in CMC—including, as we will see in section 3.C, distance education, hypertext and hypermedia, as well as some games, including stand-alone as well as MMOGs, and virtual worlds such as *Second Life*, along with ongoing work on constructing "virtual reality" through stand-alone environments such a CAVE (Cave Automatic Virtual Environment), etc. I will comment on some of these briefly, but I focus on virtual communities for the basic reason that in the context of 1990s' CMC research and popular attention, virtual communities served as the primary example and focus on "the virtual," and so they remain at the forefront for the sake of this historical overview.

   At the same time, however, as discussed in our Introduction, Johnny Søraker marks out important distinctions between virtual communities and virtual worlds (see p. 61).
2. This includes Descartes' pure mind as the *ego cogito*, the thinking self that can only be certain of its own existence, in contrast with body and the natural order more broadly as both epistemologically uncertain and ontologically inferior to the mind.
3. It may be that this strongly dualistic way of conceiving the virtual vs. the real, especially as correlated with regard to postmodernist vs. modernist conceptions of identity, further rests upon Lyotard's (1979) construal of postmodernism as the rejection of "the master narratives" of modernity, including the basic Enlightenment philosophies that fostered modern liberal democracies as legitimated by individual, autonomous selves of the sort articulated by Locke and Kant. That is to say, on this view, Lyotard's rejection of such Enlightenment philosophies *tout court* thus issues in the either/or at work here between modernist and postmodernist conceptions of the self. Insofar as this is true, it would complement my attention here to *Neuromancer* and thereby Descartes and Augustine as ultimate roots of this dualism. But to develop this side of the argument is beyond the bounds of this paper.
4. Interestingly, this apparent difficulty of forgetting our bodies in virtual environments extends to the close correspondence between our real-world identities and our avatars—including the proxemics or appropriate "body" distance human users maintain for their virtual counterparts (Consalvo, 2010, 337).
5. It may be worth noting here that McLuhan may have also been prescient in this direction with his forecast of computers and data banks as leading to an age of co-presence, an implosion in which "everybody is involved with everybody" (McLuhan and Parker 1968, p 35).
6. While the *political* dimensions of these shifts are beyond the scope of this paper, here it is worth pointing out that what David suggests as a return to a more medieval lack of individual privacy, with the modern bathroom and bedroom as the premier instantiations of such privacy, does raise a major point of concern regarding the shift towards the relational self. I have discussed how such a shift may indeed portend a (re)turn to the more authoritarian regimes historically affiliated with pre-modern and non-Western societies correlated with such relational selves (2009, 2010). Even more dramatically, Jakob Linaa Jensen (2007) has argued that the internet, as facilitating multiple sorts of surveillance, goes well beyond Foucault's well-known exposition of Jeremy Benthem's panopticon, in which one (guard) watches many (prisoners). Rather, Jensen demonstrates how the internet facilitates "everybody watching everybody" in what he calls the internet omnopticon—one which he, antic-

ipating David, likewise characterizes as entailing a return to the medieval village, but now understood from a political perspective as a regime devoid of modern notions of (individual) rights, etc.

7. Indeed, we can note that similar predictions of revolutionary change were made in conjunction with 1980s and 1990s explorations of hypertext and hypermedia (e.g., Bolter, 1984, 1991; Landow, 1992, 2006). For better and for worse, however, the more revolutionary (i.e., more dualistic) visions of early hypertext, i.e., as leading to the end of the book (literacy-print) and with it, the (modernist-individual) author, have not been realized. Rather, in ways directly parallel with the turn from more dualistic to more complementary understandings of the relations between the virtual and the real that we are examining primarily with regard to virtual communities, hypertexts and hypermedia these days, at least in the extensive and sophisticated forms envisioned by such pioneers as Bolter and Landow, remain largely at the margins of new media production and consumption, primarily through the software product *Storyspace* (which Bolter helped develop).
8. As with the early visions of hypertext and hypermedia (see previous note), and thus in ways that directly echo the sorts of Cartesian dualisms we have seen prevailing in 1990s' approaches to cyberspace and virtual reality, so proponents of distance education in those days confidently predicted the end of brick-and-mortar universities—i.e., presuming that the virtual could entirely replace and render as obsolete the real and the material. Suffice it to say here that this dualism has likewise collapsed, to be replaced with a more balanced approach that seeks to conjoin the advantages of both face-to-face education with the best potentials of online communication. See Ess (2007) for discussion.
9. The work of Michael Polanyi and his exploration of *tacit* knowledge likewise highlights the role of the body in our knowing the world—including our making the multiple *judgments* necessary in order to develop more explicit forms of knowledge, beginning with natural science (1967, 1969).
10. In philosophical terms, this discussion can begin with the thematic Maria Bakardjieva notes as a component of what she identifies as critical methodologies in the study of the internet, i.e., Frankfurt School analyses of how the market and market models encroach upon the lifeworld (see Bakardjieva, 2010, 76). It is perhaps telling that the recent research on online communities and social networking sites points to the dangers of commodification—most especially for young people and women, who remain comparatively vulnerable: in addition to Kendall's invocation of Borgmann on this point (2010, p 319), see Bromseth & Sundén, 2010, 283f; Lüders, 2100, 463; Livingstone, 2010, 354; Baym, 2010, 399. To my knowledge, however, this discussion has yet to be taken up in earnest, at least within the context of CMC research.

References

Albrechtslund, A. (2008). Online Social Networking as Participatory Surveillance. *First Monday 13* (3), March, 2008. http://firstmonday.org/article/view/2142/1949

Ames, R., and Rosemont, H. (1999). *The Analects of Confucius: A Philosophical Translation*. New York: Ballantine Books.

Bakardjieva, M. (2005). *Internet Society: The Internet in Everyday Life*. London, Thousand Oaks, New Delhi: Sage.

Bakardjieva, M. (2010). Internet in Everyday Life: Diverse Approaches. In Mia Consalvo and Charles Ess (eds.), *The Blackwell Handbook of Internet Studies*, (pp. 59–82). Oxford: Wiley-Blackwell, 2010.

Bakardjieva, M. and Gaden, G. (2010). Web 2.0 Technologies of the Self. Unpublished manuscript.

Baron, N. (2008). *Always On: Language in an Online and Mobile World.* Oxford: Oxford University Press.

Barlow, J.P. (1996). A Declaration of the Independence of Cyberspace.<http://w2.eff.org/Censorship/Internet_censorship_bills/barlow_0296.declaration>. Retrieved 14 May 2010.

Barney, D. (2004). The Vanishing Table, or Community in a World That Is No World. In A. Feenberg and D. Barney (eds.), *Community in the Digital Age: Philosophy and Practice*, (pp. 31–52). Lanham, MD: Rowman & Littlefield.

Baym, N. (1995). The Emergence of Community in Computer-Mediated Communication. In S.G. Jones (ed.), *CyberSociety: Computer-Mediated Communication and Community* (pp. 138–163). Thousand Oaks, CA: Sage.

Baym, N. (2010). Social Networks 2.0. In M. Consalvo and C. Ess (eds.), *The Blackwell Handbook of Internet Studies* (pp. 384–405). Oxford: Wiley-Blackwell.

Becker, B. (2001). The Disappearance of Materiality? In V. Lemecha and R. Stone (eds.), *The Multiple and the Mutable Subject* (pp. 58–77). Winnipeg: St. Norbert Arts Centre.

Bolter, J.D. (1984). *Turing's Man: Western Culture in the Computer Age*. Chapel Hill: University of North Carolina Press.

Bolter, J.D. (1991). *Writing Space: The Computer, Hypertext, and the History of Writing*. Hillsdale, NJ: Lawrence Erlbaum.

Bolter, J.D. (2001). Identity. In T. Swiss (ed.), *Unspun* (17–29). New York: New York University Press. Available online: <http://www.nyupress.nyu.edu/unspun/samplechap.html.>

Borgmann, A. (1984). *Technology and the Character of Contemporary Life: A Philosophical Inquiry*. Chicago: University of Chicago Press.

Borgmann, A. (1999). *Holding onto Reality: The Nature of Information at the Turn of the Millennium*. Chicago: University of Chicago Press.

Borgmann, A. (2004). Is the Internet the Solution to the Problem of Community? In A. Feenberg & D. Barney (eds.), *Community in the Digital Age* (pp. 53–67). Lanham, MD: Rowman & Littlefield.

Boyd, D. (2008). Why Youth (Heart) Social Network Sites: The Role of Networked Publics in Teenage Social Life. In D. Buckingham (ed.), *Youth, Identity and Digital Media* (pp. 119–142). Cambridge, MA: MIT Press.

Bromseth, J., and Sundén. J. (2010). Queering Internet Studies: Intersections of Gender and Sexuality. In M. Consalvo and C. Ess (eds.), *The Blackwell Handbook of Internet Studies* (pp. 270–299). Oxford: Wiley-Blackwell.

Buchanan, E. (2010). Internet Research Ethics: Past, Present, Future. In Mia Consalvo and Charles Ess, editors, *The Blackwell Handbook of Internet Studies* (pp. 83-108). Oxford: Wiley-Blackwell.

Chesebro, J.W., and Bertelsen, D.A. (1996). *Analyzing Media: Communication Technologies as Symbolic and Cognitive Systems*. New York: The Guilford Press.

Consalvo, M. (2010). MOOs to MMOs: The Internet and Virtual Worlds. In M. Consalvo and C. Ess (eds.), *The Blackwell Handbook of Internet Studies* (pp. 326–347). Oxford: Wiley-Blackwell.

David, Gabriela. (2009). Clarifying the Mysteries of an Exposed Intimacy: Another Intimate Representation *Mise-en-scéne*. In Kristüf Nyíri (ed.), *Engagement and Exposure: Mobile Communication and the Ethics of Social Networking* (pp. 77–86). Vienna: Passagen Verlag.

Dreyfus, Hubert. (1972). *What Computers Can't Do*. Cambridge, MA: MIT Press.

Dreyfus, Hubert. (1992). *What Computers Still Can't Do: A Critique of Artificial Reason*. Cambridge, MA: MIT Press.

Dreyfus, Hubert. (2001). *On the Internet*. London and New York: Routledge.

Ess, C. (1995). Reading Adam and Eve: Re-Visions of the Myth of Woman's Subordination to Man. In M. M. Fortune and C. J. Adams (eds.), *Violence Against Women and Children: A Christian Theological Sourcebook* (pp. 92–120). New York: Continuum Press.

Ess, C. (2004). Beyond *Contemptus Mundi* and Cartesian Dualism: Western Resurrection of the BodySubject and (re)New(ed) Coherencies with Eastern Approaches to Life/Death. In G. Wollfahrt and H. Georg-Moeller (eds.) *Philosophie des Todes: Death Philosophy East and West* (pp. 15–36). Munich: Chora Verlag.

Ess, C. (2005). Beyond Contemptus Mundi and Cartesian Dualism: The BodySubject, (re)New(ed) Coherencies with Eastern Approaches to Life/Death, and Internet Research Ethics. In M. Thorseth and C. Ess (eds.), *Technology in a Multicultural and Global Society*, 33–50. Programme for Applied Ethics: Publication Series No. 6. Trondheim: Norwegian University of Science and Technology.

Ess, C. (2007). Liberal Arts and Distance Education: Can Socratic Virtue (*arete*) and Confucius' Exemplary Person (*junzi*) Be Taught Online? In M. Pegrum and J. Lockard (eds.), *Brave New Classrooms: Educational Democracy and the Internet* (pp. 189–212). New York: Peter Lang.

Ess, C. (2009). Global Convergences, Political Futures? Self, Community, and Ethics in Digital Mediatized Worlds. (Inaugural lecture). Sept. 19, 2009, Aarhus University. <http://www.podcast.au.dk/?p=episode&name=2009-10-13_charles_ess_tiltrdelsesforlsning.flv> Retrieved 16 May, 2010.

Ess, C. (2010a). The Embodied Self in a Digital Age: Possibilities, Risks, and Prospects for a Pluralistic (Democratic/Liberal) Future? *Nordicom Review*. 31: 105–118. <http://www.nordicom.gu.se/common/publ_pdf/320_10%20ess.pdf>

Ess, C. (2010b). Trust and New Communication Technologies: Vicious Circles, Virtuous Circles, Possible Futures. *Knowledge, Technology, and Policy*. DOI 10.1007/s12130-010-9114-8.

Floridi, L. (2007). A Look into the Future Impact of ICT on Our Lives. *The Information Society, 23* (1): 59–64.

Foucault, M. (1988). Technologies of the Self. In L. H. Martin, H. Gutman, & P. Hutton (eds.), *Technologies of the Self: A seminar with Michel Foucault* (pp. 16–49). Amherst: The University of Massachusetts Press,

Gibson, W. (1984). *Neuromancer*. New York: Ace Books.

Goffman, E. ([1959] 1990). *The Presentation of Self in Everyday Life*. London: Penguin.

Goffman, E. (1974). *Frame Analysis*. New York: Harper & Row.

Hayles, Katherine. (1999). *How We Became Posthuman: Virtual Bodies in Cybernetics, Literature, and Informatics*. Chicago: University of Chicago Press.

Herring, S. (1996). Posting in a Different Voice: Gender and Ethics in Computer-Mediated Communication. In C. Ess (ed.), *Philosophical Perspectives on Computer-Mediated Communication* (pp. 115–145). Albany: State University of New York Press.

Herring, S. (1999). The Rhetorical Dynamics of Gender Harassment On-line. *The Information Society, 15*(3): 151–167.

Hongladarom, S., and Ess, C. (eds.) (2007). *Information Technology Ethics: Cultural Perspectives.* Hershey, PA: IGI Global,

Jensen, Jakob Linaa. (2007). The Internet Omnopticon. In H. Bang and A. Esmark (eds.), *New Publics with/out Democracy* (pp. 351–380). Copenhagen: Samfundslitteratur.

Jensen, K.B. (2010). New Media, Old Methods—Internet Methodologies and the Online/Offline Divide, in M. Consalvo and C. Ess (eds.) *The Blackwell Handbook of Internet Studies,* 43-58. Oxford: Wiley-Blackwell.

Kendall, L. (2002). *Hanging Out in the Virtual Pub: Masculinities and Relationships Online.* Berkeley: University of Calirfornia Press.

Kendall, L. (2010). Community and the Internet. In M. Consalvo and C. Ess (eds.), *The Blackwell Handbook of Internet Studies* (pp. 310–325). Oxford: Wiley-Blackwell.

Kolko, B. and Reid, E. (1998). Dissolution and Fragmentation: Problems in Online Communities. In S. Jones (ed.) *Cybersociety 2.0* (pp. 212–31). Thousand Oaks, CA: Sage.

Kolko, B., Nakamura, L., and Rodman, G.B. (eds.). (2000). *Race in Cyberspace.* London: Routledge.

Kondor, Z. (2009). Communication and the Metaphysics of Practice: Sellarsian Ethics Revisited. In Kristóf Nyíri (ed.), *Engagement and Exposure: Mobile Communication and the Ethics of Social Networking* (179–187). Vienna: Passagen Verlag.

Koskela, H. (2006). The Other Side of Surveillance: Webcams, Power and Agency. In D. Lyon (ed.), *Theorizing Surveillance: the Panopticon and Beyond,* pp. 163-181. Oxfordshire, UK: Willan.

Lanham, R. (1993). *The Electronic Word: Democracy, Technology, and the Arts.* Chicago: University of Chicago Press.

Landow, G.P. (1992). *HYPERTEXT: The Convergence of Contemporary Critical Theory and Technology.* Baltimore: The Johns Hopkins University Press.

Landow, G.P. (2006). *Hypertext 3.0: Critical Theory and New Media in an Era of Globalization.* Baltimore: The Johns Hopkins University Press.

Livingstone, S. (2010). Internet, Children, and Youth. In M. Consalvo and C. Ess (eds.), *The Blackwell Handbook of Internet Studies* (pp. 348–368). Oxford: Wiley-Blackwell.

Lyotard, J.F. (1979). *The Postmodern Condition.* Minneapolis: University of Minnesota Press.

Lüders, M. (2007). *Being in Mediated Spaces: An Enquiry Into Personal Media Practices.* (Ph.D. thesis, University of Oslo).

Lüders, M. (2010). Why and How Online Sociability Became Part and Parcel of Teenage Life. In M. Consalvo and C. Ess (eds.), *The Blackwell Handbook of Internet Studies* (pp. 456–473). Oxford: Wiley-Blackwell.

Løgstrup, K.E. (1956). *Den Etiske Fordring* [The Ethical Demand]. Gyldendal: Copenhagen.

McLuhan, M., and Parker, H. (1968). *Through the Vanishing Point: Space in Poetry and Painting.* New York: Harper.

Myskja, B.K. (2008). The Categorical Imperative and the Ethics of Trust. *Ethics and Information Technology, 10*: 213–220.

Negroponte, N. (1995). *Being Digital.* New York: Knopf.

Nikamura, L. (2000). Where Do You Want to Go Today, in Beth Kolko, Lisa Nakamura, and Gilbert B. Rodman (eds.), *Race in Cyberspace,* pp. 15–26. New York: Routledge.

O'Leary, S.D., & Brasher, B. (1996). The Unknown God of the Internet. In C. Ess (ed.), *Philosophical Perspectives on Computer-Mediated Communication* (pp. 233–269). Albany: State University of New York Press.

Polanyi, M. (1967). *The Tacit Dimension.* Chicago: University of Chicago Press.

Polanyi, M. (1969). *Knowing and Being.* Chicago: University of Chicago Press.

Reading, A. (2009). The Playful Panopticon? Ethics and the Coded Self in Social Networking Sites. In Kristóf Nyíri (ed.), *Engagement and Exposure: Mobile Communication and the Ethics of Social Networking* (pp. 93–101). Vienna: Passagen Verlag.

Rheingold, H. (1993). *The Virtual Community: Finding Connection in a Computerized World.* Boston, MA: Addison-Wesley Longman.

Rheingold, H. (2000). *The Virtual Community: Homesteading on the Electronic Frontier.* Cambridge, MA: MIT Press.

Röttgers, K (2009). The Pornographic Turn, Or, the Loss of Decency. In Kristóf Nyíri (ed.), *Engagement and Exposure: Mobile Communication and the Ethics of Social Networking* (pp. 87–91). Vienna: Passagen Verlag.

Stone, A. R. (1991). Will the Real Body Please Stand Up? Boundary Stories about Virtual Cultures. In M. Benedikt (ed.), *Cyberspace: First Steps* (pp. 81–118). Cambridge, MA: MIT Press.

Stuart, S. (2008) From agency to apperception: Through Kinaesthesia to Cognition and Creation. *Ethics and Information Technology 10* (4): 255–264.

Turkle, S. (1995). *Life on the Screen: Identity in the Age of the Internet.* New York: Simon & Schuster.

Storyspace (hypermedia software). Watertown, Maine: Eastgate Systems. <http://www.eastgate.com/storyspace/index.html>

Vallor, S. (2009). Social Networking Technology and the Virtues. *Ethics and Information Technology.* DOI 10.1007/s10676-009-9202-1.

Van Gelder, L. ([1985] 1991). The Strange Case of the Electronic Lover. In C. Dunlop and R. Kling (eds.), *Computerization and Controversy* (pp. 364–375). San Diego: Academic Press, 1991. (Originally published in: *Ms. Magazine,* October 1985, pp. 94–124.)

Wellman, B., and Haythornthwaite, C. (2002). *The Internet in Everyday Life.* Oxford, Malden, MA: Blackwell.

CHAPTER TWO

'Virtual Reality' and 'Virtual Actuality'

Remarks on the Use of Technical Terms in Philosophy of Virtuality

MARIANNE RICHTER

> Philosophy is the art of forming, inventing and fabricating concepts.
>
> —*DELEUZE & GUATTARI*[1]

Introduction

The goal of this essay is to reveal (1) important etymological roots and (2) conceptual differences between 'reality', 'actuality' and 'virtuality' and (3) to shed light on the occurrence of these terms in philosophy of virtuality. It is intended to provide an overview therewith that allows further investigations into the logical status[2] of the respective terms. This overview, however, should be taken merely as a starting point for a more in-depth investigation. Even so, it shall help to distinguish exemplary ways of using the terms as technical terms with regard to syntactical as well as semantic differences. By pointing out not only manifold but also incoherent concepts of especially 'virtuality' I would like to emphasize the need for the more careful explication of the conceptual and terminological references of these terms, particularly in multidisciplinary contexts.

It should be convenient to begin with the explication of some general assumptions underlying the attempt to investigate different uses of technical terms, for the adequacy of their use can of course be regarded as depending on context. It would seem that criteria of adequacy differ synchronically as well as diachronically while any attempt to implement a certain usage could be taken as an imperial purpose—a step backwards according to postmodernist rejections of regulations and standards, a purpose that reveals unawareness of the frequent (and perhaps fruitful) development of terminologies.

Still, the plea for an adequate, that is, at least *coherent,* use of technical terms seems to be justifiable with regard to a pragmatic argument: in terms of inter-subjective exchange and development of ideas, it certainly is helpful to refer to a common conceptual and terminological basis. Therefore, I take it as an initial requirement that scientific discourses are ideally characterized by a coherent use of technical terms. I do *not* insist that concepts can be construed in a definite way, brought into one consistent order and/or that there is a need for determining the 'right' or the original way of using a technical term with regard to the history of concepts.

Concerning the relevance of research on the history of concepts, I rather agree with Hofmann, who argues that research of this kind brings classical, conventional or canonical concepts back into critical discussions by revealing further (perhaps even forgotten) aspects of key terms and ideas (Hofmann, 2002, p. 149). In addition I would like to emphasize that the history of concepts might also suggest further ways of comparing and relating different concepts and terms with one another.

This briefly indicates the relevance of the goal introduced above. Since philosophy of virtuality is a relatively young branch of philosophy, I would like to clarify existing as well as emerging conventions for using technical terms (especially 'virtual reality' and 'virtual actuality') as a first, modest step towards a coherent discourse—at least insofar as this analysis makes clear for us important incoherencies in the discourse under consideration.

Etymological roots and concepts of 'reality'

Realitas, the Latin root of 'reality' by common consent, probably stems from the term *res* ('thing'). Accordingly, *realitas* first referred to the *essentia* ('nature') or *essentialitas* ('essentiality') of *res* ('thing') (Courtine, 1992a, p. 178). In this context, *realitas* presumably denoted the thing-quality or thing-ness of things in general (Courtine, 1992a, p. 178). If so, *realitas* was already an abstract (in this case property-denoting) concept in its primary scholastic usage, which is worth keeping in mind with regard to further concepts of reality outlined below.

Moving beyond the Middle Ages (which might be worth a particular examination in its own right), it is illuminating to focus on significant differences between *realitas objectiva* (Descartes) and *objektive Realität* (Kant), since especially Kant is regarded as having had a significant influence on contemporary concepts of reality (Courtine, 1992b, p. 189). By all means, the following comparison tentatively suggests some conceptual variations of the primary concept of *realitas* as briefly noted above.

In *Meditationes de Prima Philosophia*, Descartes distinguishes between *realitas objectiva* and *realitas actualis/realitas formalis* (Med. 3, p. 32ff.).[3] This distinction seems to result from different views on thoughts: either one can focus on the mode of thinking-a-thought that all thoughts have in common—or one can consider the representational content of thoughts. Accordingly, the representational contents of thought (e.g., a red rose) share their being-a-thought (which is their *realitas actualis*), but differ in their being-a-thought-of-something (which is their *realitas objectiva*). Despite the fact that the term *realitas* mainly occurs in certain combinations, it could be suggested to rephrase it as being-ness according to the aforementioned combinations.

Kant's use of *objektive Realität* is different from Descartes' in the way that he distinguishes between *objektive Realität* ('objective reality') and *Idealität* ('ideality'). The predicative use of *objektive Realität* can be construed as indicating that one can intentionally refer to a certain object and, moreover, that this object is given in sensory experience (KrV, B, 147–150). By contrast, the term 'ideality' is used to indicate that there is an intentional referent, but no intuitional exemplification (KrV, B, 44). To offer a simple explanation of this: something can be attributed as 'objectively real' in the Kantian sense when one can literally point to it, whereas 'ideality' signifies that there is no possibility of pointing to a corresponding referent in intuition. It is important, after all, that *objektive Realität* obviously remains an abstract concept here: though it neither names the thing-quality or things-ness of things nor the factual content of thoughts in general, it refers to the notion of experienceability as such (by all means not to a certain cluster of things outside of mind).

Today the meaning of 'reality' ranges between several alternatives. Two of them shall be introduced as examples. C. Hubig, for instance, regards reality as referring to the whole of acknowledged facts, while *Wirklichkeit*[4] refers to the whole of effects on us as subjects (Hubig, 2006, p. 187ff.). This distinction apparently aims to capture the difference between a linguistically composed and theoretically grounded worldview, which would be a specific concept of reality, and experiences given within the present moment that might fit or confront a worldview. As a matter of fact, this way of using the term *Wirklichkeit*, namely assigning it to the notion of instant experience, is often traced back to Meister Eckhart, who translated the Latin *actualitas* with *Wirklichkeit* (Sachsse, 1974, p. 11f.).

Still, a quite different use of 'reality' is current and preferred for some reasons: specifically, reality is considered to refer to what is regarded as factually given, in contrast with the German *Wirklichkeit*, as meaning the comprehensive system of the factually given, including all those symbols and principles that provide systemizing functions (Wiegerling, 2010). This usage of the terms aims to confront the aforementioned one for several reasons. As K. Wiegerling (2010) explains, first, the use of *Wirklichkeit* inspired by the medieval precedents carries with it metaphysical notions that are irreconcilable with significant epistemological attainments of the 19th century. Given 19th-century idealism—i.e., as precluding our access to the 'factual' without some sort of shaping schema (such as Kant's frameworks of time and space, the categories of understanding, etc.)—then referring to *Wirklichkeit* as meaning something beyond or opposed to the 'factual' that affects us in an 'unfiltered' or 'immediate' way is no longer meaningful. Second and in addition to the first argument: we have an idea of what it means to (temporarily) loose one's 'reality', but it is not at all clear what a loss of *Wirklichkeit* or 'actuality' could mean (Wiegerling, 2010). Current worldviews (e.g., what I consider as being the fact right now) might differ under the effect of medications or perhaps under the constant influence of virtuality. But only if we lost our capability to formulate what (actually or principally) appears to be the fact, *Wirklichkeit* would be in danger. Hence, the identification of 'reality' with the factually given seems much closer to the current linguistic conventions and therefore more appropriate (given, of course, that technical terms should correspond to linguistic conventions).

However, while trying to avoid such backward steps in metaphysics, recent concepts of reality likely forfeit a notion that (once) has been related with this term. As R. Rorty (2007) argues, '[i]f the word reality were used simply as a name for the aggregate of all [justified and non-justified beliefs about] such things, no problem about access to it could have arisen. The word would never have been capitalized' (p. 105f.). According to Rorty, epistemological doubts primarily arise from the assumption of a 'real reality', which is hidden behind the surface of what is considered as 'reality'. However, judgments about the 'real reality' tend to be considered as unjustifiable: 'But when [reality] is given the sense that Parmenides and Plato gave it, nobody could say what would count as a justification for a belief about the thing denoted by the term' (Rorty, 2007, p. 105). Accordingly, 'real reality' on the one hand longs for being comprehended, but on the other hand refuses to be identifiable (e.g. via perception or discursive methods). In view of this, recent conceptions of reality might prefer the avoidance of such an indefinable extension in order to keep the corresponding term meaningful. But, as Rorty rightly notes, the core of epistemological interest is then no longer addressed by the term itself.

While this overview may be somewhat cursory, the following aspects can be regarded as significant for contemporary discussions about reality: first, the emphasis on the comprehensive extension of 'reality' (instead of focusing on the conditions for the assignment of the predicate 'real'); second, the specific ontological status of reality (in some cases it refers not only to an abstract property or idea of the whole of things that are assumed to be given factually or fictionally and/or virtually but to specific worldviews); third, the possibility of using at least one of the nouns ('reality', 'actuality') in the plural form, which apparently is a consequence of first and second; and lastly, the declining usage of the metaphysical notion of a 'real reality', which requires comparison forms such as 'more real' or 'really real.'

Etymological roots and concepts of 'actuality'

In contrast to 'reality', the term 'actuality' is only rarely found in etymological encyclopedias. But knowing that the etymological roots go back to the Latin *actus* (vs. *potens*; Schlüter, 1971), a trace to the concept of actuality can be found in Aristotle's work and especially in the scholastic translations. The Aristotelian counterparts to *actualitas* are *energeia* (i.e., the process of actualization) and *entelechia* (i.e., the state of actualization; a difference that got lost in the Latin translation).

The ambitious attempt to construe the concept(s) of actuality, as established by Aristotle, leads first of all to the opposition between actuality and potency. This opposition is crucial in the way that definitions of actuality refer to potency and vice versa. Accordingly, something actual is no longer potential but certainly has been potential before becoming actual, while something potential is not yet actual but might become actual under certain circumstances. Since potencies, however, could as well be regarded as existing or being the fact ('It *is* possible that . . .'), it has been proposed to construe actuality and potency as coexisting *modes* of being (Berti, 1996, p. 289).

Apart from the controversial relation between both concepts, the corresponding theoretical framework, developed in the ninth book of Aristotle's *Metaphysics*, raises a couple of further classical problems—for example, the epistemic problem of judging whether something is definitely potential or possible. For the purpose of this investigation, however, it is convenient to concentrate on the (assumed) intension and extension of the terms. In this regard, it is striking that neither the scope of actual nor possible entities is referred to as 'reality' or 'actuality' in Aristotle's terms. Therefore, his work hardly serves as a reference when it comes to tracing hypostatizing usages of the former predicate (Blumenberg, 1974, 3ff.).

A significant trace to the concept of actuality, as construed above, can be found in Kant's work (KpV, 39): He introduces the term *Wirklichmachung*, which refers to the notion of bringing a concern of practical reason into existence via a corresponding action. However, this term (which is no longer in use) was soon replaced with *Realisierung* ('realization') (Büttemeyer, 1992).

The substitution possibly lead to the synonymous use of 'reality' and 'actuality' as is common in daily language as well as in some philosophical works. The habit of defining one term through the other might be a significant result of this: according to several dictionaries 'actual' means 'real' and the other way round. However, the history of the concepts as outlined in brief suggests further possibilities of differentiating between these terms. First, 'actuality' is often classified as a modal term, while 'reality' is rather used as a generic term. Second, 'actuality' is opposed to 'potency', while 'reality' has no opponent of equal importance.

'Reality' and 'actuality' in philosophy of virtuality

In view of the concepts of reality and actuality, introduced above, a further concern is: if and where virtuality could be located in the corresponding theoretical frameworks. Does virtuality refer to an alternative mode of being, to certain technically induced augmentations of intuitions, and/or to something completely different? Moreover, does it make sense to use combinations such as 'virtual reality' and 'virtual actuality' and in what respect do the 'hybrids'[5] differ ontologically from 'normal' reality and actuality? At this point, the relevance of a closer look at current as well as previous concepts should be apparent, for it might countervail a conceptual confusion.

Indicating the need of conceptual clarifications, Michael Heim begins with pointing out the frequent though unspecific use of 'virtual':

> Today, we call many things 'virtual'. Virtual corporations connect teams of workers located across the country. In leisure time, people form clubs based on shared interests in politics or music, without ever meeting face-to-face. Even virtual romances flourish through electronic mail. (Heim, 1998, p. 3)

Accordingly, 'virtual' occurs in attributions (e.g., virtual romance), names (e.g. 'Virtual Corporation') or as a byword for computer-generated institutions (e.g., clubs) or computer-mediated social interactions (e.g., sharing interests). The semantic indistinctness in daily language, however, challenges each attempt to abide by a terminological use and requires even more carefulness in case of combinatory usages such as 'virtual reality' and 'virtual actuality'.

Focusing on the philosophical issues that arise in view of the broad extension(s) of the term(s), I would like to begin with the exposition and discussion of an early concept of virtual reality that was proposed in the 1990s, when various visions concerning 'virtual worlds' appeared all over the 'real world'.

P. Milgram and F. Kishino proposed the taxonomy of 'mixed realities' in 1994. The objects of interest have in particular been 'technologies that involve the merging of real and virtual worlds somewhere along the "virtuality continuum", which connects completely real environments to completely virtual ones' (Milgram & Kishino, 1994, p. 2). Obviously, Milgram and Kishino neither require an unbridgeable gap between reality and virtuality nor a third category such as virtual actuality. Instead, they distinguish between real entities and virtual entities, while the conceptual difference between virtual and real would be that *virtual* refers to what is artificial and not necessarily bound to physical laws ('existing in effect but not formally or actually'; Milgram & Kishino, 1994. p. 3) vs. the non-artificial as necessarily bound to physical laws ('having an actual objective appearance'; Milgram & Kishino, 1994, p. 3). Accordingly, the corresponding nouns 'reality' and 'virtuality' refer to different user environments ('worlds') that consist only of one of the respective types of entities, while 'mixed reality' refers to a more or less hybrid user environment.

Milgram and Kishino thus draw an opposition between 'virtual' and 'real' which is inspired from the conventional distinction between technically generated and naturally originated entities. This distinction is problematic insofar as life is obviously technically penetrated:

> Our culture intentionally fuses—sometimes even confuses—the artificial with the real, and the fabricated with the natural. As a result, we tend to quickly gloss over the precise meaning of virtual reality and apply the term virtual to many experiences of contemporary life. (Heim, 1998, p. 4)

Consequently, the distinction between technically generated and naturally originated entities possibly leads to the conclusion that everything is or has become virtual. But if everything is virtual, any initial distinction between 'real' and 'virtual' thus disappears—giving us a *reductio ad absurdum* argument against the distinction as initially defined.

Another critical point is the link between the concept of virtuality and certain technical tools that are able to synthesize virtuality. The aforementioned approach is an example of the tendency to link the meaning of virtuality with certain technologies, particularly in computer science (Heim, 1998, p. 3ff.). However, definitions of that kind are likely circular insofar as the identification of virtuality depends on the identification of technologies that create virtuality. This identification in turn requires a definition of virtuality and so forth.

Focusing on the virtual environment, one could try to clarify the notion of virtuality with an additional specification of experiences or events that are associated with it. Such experiences can include multiple aspects from 'driver training to computer-aided tourism' (Heim, 1998, p. 6). However, the problem of circularity arises here as well, since the list of inducible experiences or interactions with virtuality is underdetermined due to the constant development of technologies. That is, since the technologies are constantly changing, the list of experiences that might constitute an extensive definition of 'virtual' always lags behind and remains incomplete. Therefore, it needs to be actualized every now and then according to the special characteristics of virtuality, which are lacking insofar as the listing of experiences should have been compiled in accordance with them.

It remains to abstract specific qualities of the above-mentioned experiences in order to gain criteria which would help us define 'virtuality' more precisely, which is quite popular in the end. M. Heim (1998), for example, suggests the following 'three I's' as the fundamental characteristics of experiences in virtual reality:

> [...] Immersion, interactivity, and information intensity. Immersion comes from devices that isolate the senses sufficiently to make a person feel transported to another place. Interaction comes from the computer's lightning ability to change the scene's point-of-view as fast as the human organism can alter its physical position and perspective. Information intensity is the notion that a virtual world can offer special qualities like telepresence and artificial entities that show a certain degree of intelligent behavior. Constantly updated information supports the immersion and interactivity, and to rapidly update the information, computers are essential. (p. 7)

The difficulty is to exhibit the distinguishing marks of virtual experiences not only in terms of quantitative but also qualitative criteria, which seems hard to tackle.

In response to subsequent efforts to provide such criteria, a further taxonomy has been proposed recently. C. Hubig (2006) develops a cross-classificatory distinction between 'reality', 'actuality', 'virtual reality' and 'virtual actuality'. While 'reality' refers to the whole of acknowledged facts (including possiblities), 'actuality' refers to the whole of concrete effects on us as subjects. In accordance with that, two further concepts are proposed: 'Virtual reality', denoting the whole of simulated facts, and 'virtual actuality', denoting effects on us as subjects that are induced in confrontation with simulated facts. Philosophy of science thus deals with the realm of virtual reality insofar as it asks, Which simulated fact has or will have an actual reference? While philosophy of technology and/or ethics focuses on virtual actualities insofar as it asks, What effects and impacts are probable in the realm of modern computational technology as well as in the specific cases that are simulated (e.g., an emergency landing; Hubig, 2006, p. 187ff.)?

As for the definition of 'virtuality', the notion of being a technologically induced phenomenon remains essential (and controversial) here. Concerning the more fine-grained distinction between virtual reality and virtual actuality, the key to comprehension is to consider the term 'virtual' as always referring to the simulated facts and not to the effects. Virtual actuality thus differs from 'normal' actuality only with regard to its specific 'origin', but not with regard to its ontological status, since an effect remains an effect independent of its cause. As for the relation between reality and virtual reality, it is characterized by the subordination of virtual reality to reality. Hence, virtual reality, as being the whole of facts that are not regarded as being independent of the simultaneous use of simulation technology, is part of reality. Nonetheless, when it comes to the demand for a more precise ontological specification of virtual reality, the above-mentioned problem of circularity arises here as well.

To draw another interim conclusion: First, the conceptual opposition between reality/parts of reality and virtuality is obviously crucial in each of the exemplary approaches. The further distinction between actuality and virtual actuality, however, only appears to be ontological, but rather has a pragmatic function, since it allows for an explicit focus on technologically induced impacts. Second, virtuality tends to be regarded as something technologically induced, in contrast with reality/parts of reality. Third, the distinction between respective fabrications and other fabrications is drawn with regard to specific qualities of the facilitated states, events or experiences (e.g., immersion, interactivity, information intensity; Heim, 1998, p. 7).

The logical status of 'virtual/virtuality'

Terminologies and their corresponding theoretical backgrounds certainly do not allow for every conceivable modification. An augmentation of the Kantian concept of objective reality through a further distinction between virtual reality and reality, for example, can lead to systematic inconsistencies. On the other hand, especially the ontological approaches towards virtuality seem to be characterized by numerous borrowings from various philosophical conceptions. This, indeed, is not surprising, since it is entirely common to begin with acknowledged concepts in order to approach towards as yet un-explicated phenomena. As a result, however, not only the references to the history of concepts are manifold, but also the conceptions of this history itself.

In addition to current investigations into the various documentations of virtual and virtual reality conceptualizations, it stands to reason to ask whether this diversity hints at both a heterogeneous and incoherent use of 'virtuality' as a technical term. If the latter is the case, it should be even more necessary to avoid the suggestion of a common sense notion of virtuality.

With respect to the logical status of the term 'virtuality' it can be, indeed, shown that there are incoherencies in the use of it. The following examples should help to make this case. E. Esposito (1998), to begin with, distinguishes, between 'simulation' and 'virtuality' as different types of fiction: When fiction can generally be regarded as something made up, then there would be a difference between simulation (making up something that adapts to reality as closely as possible) and virtualization (making up a sphere of pure 'possible-possibilities'). Accordingly, Esposito suggests 'virtuality' to be a modal term, as referring to the pure thinkable, among the possible, the necessary and the actual.

In this way, Esposito sticks to what has been shown as typical for approaches towards virtuality in the previous section, namely, stressing the notion of something made-up (technically fabricated in the broadest sense) as well as the opposition to reality (since the pure-possible transcends reality in suggesting what else or instead could be thought of as being the fact). It may be, however, that Esposito may need to refrain from the combination 'virtual actuality', since that would possibly imply an iteration of modalities.[6]

A methodologically similar but conceptually different approach would be to link virtuality to the Kantian notion of ideality.[7] N. Rescher (2000) suggests a corresponding usage of the terms, when he takes up the Kantian concept of ideality:

> What Kant develops [...] is a pragmatically validated system of virtual reality—a system whose 'objects' are not real objects but ideal thought-objects (*Gedankendinge*). What we thus find in Kant is a dualistic ontology. On the one hand, there is 'the real world'—a realm of physical reality of spatio-temporal objects conceptualized by the understanding on the basis of the deliverances of sensibility. And on the other hand there is a realm of virtual reality, of intellectual quasi-objects constructed by reason in the service of its mission of systematizing deliverances of the understanding. (pp. 283f.)

It appears that Rescher focuses on the epistemic value of virtuality. This, of course, is a necessity insofar as he considers virtuality as an appropriate substitute of the Kantian concept of ideality. However, when virtual refers to ideal or the other way round, as Rescher suggests, there arise at least two questions. One, how could one explain that sensory (esp. visual) experience is assumed to be possible in terms of virtuality? Furthermore, why do users of the Internet believe that they share a virtual space and regard themselves as being able to interact within it? The last notions would probably be compatible with Rescher's in the broadest sense that we influence one another intellectually. Nevertheless, Rescher would have to refrain from the idea that ideality is a medium for social networking. Whatever the significance of this critique for Rescher's use of terms is, it further highlights the 'public' or 'inter-subjective' aspect of virtuality that also appears to be important in the discourse from the very beginning.

In view of the approaches and their critiques, as developed above, there seems to remain the possibility of taking virtuality as a sort of an exterior or externalized ideality, a whole of made-up facts available for and creatable by everyone. In connection with such a broad intersection between virtuality and reality, J. Baudrillard's (1994) approach is certainly an extreme example in that he claims no (more) ontological difference between virtuality and reality. However, when everything becomes virtual or simulated, then it no longer makes sense to establish a technical term, since technical terms principally draw distinctions. However, relying on a more or less ambiguous idea of a distinction is not at all satisfactory: *So what is virtuality anyway?* At least some characteristics of different approaches towards virtuality can be summarized now. First, similar to the concept of fiction, virtuality is defined with regard to its generation as well as with regard to the complementary product or effect ('virtuality'). Second, virtuality seems to be opposed to reality or to actuality. However, when virtual either means ideal or possible (Esposito [1998] even suggests 'only-ideal-possibilities' as to be virtual), then it is not possible to resolve *one* logical status of the term 'virtuality', since both uses—as a modal term and a generic term—are proposed.

Conclusion

My review of the history of concepts has made clear that 'actuality' has been and still is often used as a modal term, while 'reality' rather is used as a generic term. Concerning the definition of 'virtual', there are several notable tendencies: Like fiction, 'virtuality' is defined with regard to its generation (making up something, simulating something) as well as with regard to the corresponding product or complementary effect (e.g., fictional world, virtual world). One of these components is likely to be opposed to reality or actuality, so that definitions typically start from opposite notions: accordingly, 'virtual' qualifies something that *is not* real or actual in the sense of made-up in another way, appearing differently or referring to a different mode of being. But when 'virtual' thus denotes either 'ideal' or 'possible' (again Esposito [1998] suggests 'only-ideal-possibilities' as to be virtual), then there is not one logical status proposed, but two, that is, the term is used as a modal term and as a generic term under the common umbrella of philosophy of virtuality. Insofar as it is not possible to use a term as a modal term and a generic term simultaneously, one can claim that there is, indeed, a significant incoherency in the current use of 'virtuality' as a technical term. This incoherency might be important with respect to combinations such as 'virtual reality' and 'virtual actuality', since the latter, for example, might involve an iteration of modalities.

Finally, the question is how to cope with the exposed diversity of concepts hidden behind the frequently used word 'virtuality'. On the one hand, one could clarify the commonalities and peculiarities of different usages in order to provide a point of reference for further discussions. On the other hand, conceptual arrangements likely constrict the dynamics of developing concepts (Gadamer, 1972, p. 392). At last, one might come to the conclusion that the diversity of concepts and terms itself is not philosophically problematic, but the risk of forgetting about this diversity certainly is.

Acknowledgments

I would like to thank the German Research Foundation (DFG) for financial support of the *Cluster of Excellence Simulation Technology* (EXC 310/1) project at the University of Stuttgart. I am also very grateful to the participants of the International PhD/Graduate course and workshop *Philosophy of Virtuality,* held at the NTNU, 2009, and to the participants of the graduate colloquium at the institute of philosophy (Stuttgart) for inspiring comments and fruitful discussions.

Notes

1. Deleuze & Guattari, 1994, p. 2.
2. 'Logical status' refers to the way in which we use a term, e.g., as a modal term or as a generic term. I consider generic terms as super-ordinate concepts that allow for the distinction between sub-concepts (e.g., 'elephant' and 'dolphin' can be subsumed under the generic term 'animal'). Modal terms refer to a different usage of terms in the way that they denote modalities of facts or propositions. Basic modal terms are 'necessary', 'possible' and 'actual'.
3. Descartes later deviated from these highly criticized terms. But since I emphasize the systematic comparison of concepts, this withdrawal is not of special importance here.
4. There is an ongoing debate about translating *wirklich/Wirklichkeit* into English. The terms 'actual'/'actuality' have been proposed for this, but some regard this translation as inadequate (Wiegerling, 2010).
5. The combination 'virtual reality' was coined by J. Lanier in 1986 (Heim, 1998, p. 5). The combination 'virtual actuality' was introduced two decades later (Hubig, 2006, pp. 187 ff.).
6. This would particularly be the case if 'virtual' denoted the mode of being possibly possible and 'actual' (in accordance with Aristotle) denoted the mode of being in existence in contrast to being (just) possible. Sure enough, threefold iterations (such as the actual-possible-possible) might occur in logical calculations but their use in contexts like above=mentioned one would need of an explanation concerning the relevance.

7. Insofar as both approaches suggest the renaming of a certain concept: While Esposito links virtuality to contingency, Rescher uses the term 'virtual reality' to clarify the Kantian notion of 'ideality'.

References

Aristotle. (1995). *Metaphysik*. (H. Bonitz, Trans.) Hamburg: Meiner.

Baudrillard, J. (1994). Die Simulation. W. Welsch (Ed.), *Wege aus der Moderne. Schlüsseltexte der Postmoderne—Diskussion* (pp. 153–162). (2nd ed.). Berlin: Akademie-Verlag.

Berti, E. (1996). Der Begriff der Wirklichkeit in der Metaphysik. C. Rapp (Ed.), *Aristoteles: Metaphysik, Die Substanzbücher*. Berlin: Akademie Verlag.

Blumenberg, H. (1974). Vorbemerkungen zum Wirklichkeitsbegriff. G. Bandman, H. Blumenberg, H. Sachsse, H. Vormweg and D. Wellershoff (Eds.), *Zum Wirklichkeitsbegriff* (pp. 3–10). Mainz: Akademie der Wissenschaften und der Literatur.

Büttemeyer, W. (1992). Realisierung/Realisation. In *Historisches Wörterbuch der Philosophie* (Vol. 8, pp. 143–146). Basel: Historisches Wörterbuch der Philosophie.

Courtine, J.-F. (1992a). Realitas. In *Historisches Wörterbuch der Philosophie* (Vol. 8, pp. 178–185). Basel: Historisches Wörterbuch der Philosophie.

Courtine, J.-F. (1992b). Realität/Idealität. In *Historisches Wörterbuch der Philosophie* (Vol. 8, pp. 185–193). Basel: Historisches Wörterbuch der Philosophie.

Deleuze, G., and Guattari, F. (1994). *What is philosophy?* New York: Columbia University Press.

Descartes, R. (1954). *Meditationen* [Med]. (A. Buchenau, Trans.)Hamburg: Meiner.

Esposito, E. (1998). Fiktion und Virtualität. S. Krämer (Ed.), *Medien, Computer, Realität: Wirklichkeitsvorstellungen und Neue Medien* (pp. 269–296). Frankfurt a. Main: Suhrkamp.

Gadamer, H.-G. (1972). *Wahrheit und Methode*. (3rd ed.). Tübingen: Mohr.

Heim, M. (1998). *Virtual realism*. New York: Oxford University Press.

Hoffmann, S. (2002). *Geschichte des Medienbegriffs*. Hamburg: Meiner.

Hubig, C. (2006). *Die Kunst des Möglichen I*. Bielefeld: Transcript.

Kant, I. (1998). *Kritik der reinen Vernunft* [KrV]. Hamburg: Meiner.

Kant, I. (2003). *Kritik der praktischen Vernunft* [KpV]. Hamburg: Meiner

Milgram, P., & Kishino, F. (1994). A taxonomy of mixed reality visual displays. *IEICE Transactions on Information Systems*, Vol E77-D No.12. Retrieved from: http://vered.rose.utoronto.ca/people/paul_dir/IEICE94/ieice.html.

Rescher, N. (2000). Kant on the limits and prospects of philosophy—Kant, pragmatism, and the metaphysics of virtual reality. *Kant-Studien, 91*, 283—328.

Rorty, R. (2007). Pragmatism and romanticism. R. Rorty (Ed.), *Philosophy as cultural politics* (pp. 105–120). Cambridge: Cambridge University Press.

Sachsse, H. (1974). Naturwissenschaft, Technik und Wirklichkeit. G. Bandman, H. Blumenberg, H. Sachsse, H. Vormweg and D. Wellershoff (Eds.), *Zum Wirklichkeitsbegriff* (pp. 11–20). Mainz: Akademie der Wissenschaften und der Literatur.

Schlüter, D. (1971). Akt/Potenz. In *Historisches Wörterbuch der Philosophie* (Vol. 1, pp. 134–142). Basel: Historisches Wörterbuch der Philosophie.

Wiegerling, K. (2010). *Philosophie intelligenter Welten*. Forthcoming.

CHAPTER THREE

Virtual Entities, Environments, Worlds and Reality

Suggested Definitions and Taxonomy

JOHNNY HARTZ SØRAKER

Introduction

In the current literature on 'virtuality' and related terms, there is hardly any consensus on what these terms should refer to, nor what their defining characteristics are. At one end of the spectrum, computer scientists often use the term to denote a method of simulating a piece of computer hardware by other means, for instance '*virtual* memory'. At the other end of the spectrum, we find science-fiction tales of humans immersed in environments that simulate the actual world in such richness and accuracy that the two cannot be distinguished. In between, we find more mundane uses of the term, as applied to computer games and Web sites. This conceptual muddle is unfortunate, especially when our goal is to evaluate their societal and ethical impact. For instance, if we ask the question of how far trust may be established in virtual environments, our answers will (or *should*) differ depending on whether we mean 'virtual environment' to refer to chat rooms, discussion forums, email, MMOs, text messaging, and so forth. The purpose of this paper is to distinguish between four related yet fundamentally different categories of virtuality: 'virtual' (generic), 'virtual environment', 'virtual world', and 'virtual reality' and to show how they give rise to different sets of considerations when evaluating their potential impact on their users.

First of all, as if referring to virtual entities is not difficult enough, the seemingly vanishing line between real and virtual prompts us to be precise about what is 'real' as well. As I will return to below, many virtual states of affairs are just as real as (or no less illusory than) 'real reality'. Thus, I agree with Albert Borgmann's claim that 'a distinctive term for real reality needs to be found now that reality can be virtual' (Borgmann, 1999, p. 256).[1] Following his suggestion, I will use 'actual reality' when I refer to what is often misleadingly characterized as 'the real world'. I will sometimes use 'physical reality' when I explicitly address the physical as opposed to non-physical aspects of reality. As Borgmann does, I will sometimes use 'real' and 'the real thing' when there is no danger of confusion or when I address claims that specifically invoke these terms.

'Virtual' is often defined along the same lines as 'quasi-', 'pseudo-' or 'almost the same as'; something that is almost but not quite real, or something real without being *actual* (cf. Shields, 2003, p. 25, inspired by Proust and Deleuze). This usage is reflected in our daily language when we speak of something being *virtually* something else. In the same vein, Jaron Lanier, one of the early pioneers in virtual reality research who also popularized the term, explains that for something to be virtual 'it has to be indistinguishable [from the actual entity] in some practical context, while it remains distinguishable in another' (Lanier, 1999). Although this description gives us some idea what virtuality refers to and the notion of 'similar but not equal' will later prove important, the description is too imprecise and inclusive. First, it does not single out which properties of the entity in question are not actual. For instance, the term 'virtual soldier' can be applied to someone who is a human being and almost a soldier, or it can refer to 'someone' who is not a physical being, but has the properties of an actual soldier (e.g., a video game character). Second, if we admit anything that is 'almost but not quite real' in our definition, this seems to entail that vivid memories, dreams, movies and rainbows are virtual as well—which they are not, as I will return to shortly. As these examples show, a definition along the lines of 'almost real' fails to signify both how 'virtual' differs from 'actual' and, just as importantly, how it differs from thought constructs, dreams and illusions. Still, it is important to keep this aspect in our definition, because many of the philosophical issues arise precisely because our theoretical and ethical frameworks become difficult to apply when only *some* of a virtual entity's properties are actual. The challenge lies in saving the intuition that lies behind virtuality as something 'almost real', while avoiding the lack of precision that often follows.

My first step will be to define virtual as a generic term, but when I exclude certain criteria as necessary for such a broad conception of virtuality, I will return to some of these when outlining the different sub-categories. A promising starting point is to look at how 'virtual' relates to what is often referred to as 'new media'—and why it is so widespread to employ spatial metaphors.

Virtual as new media—the 'place' metaphor

There are two features that are often attributed to new media: instant, global communication, and digitized information. However, these properties are not unique to new media. First, different forms of telecommunication have been used for long-distance, instant communication since the early 20th century. Although computer-mediated communication further improves the availability, speed and reliability of long-distance communication, this does not in itself constitute a qualitative shift from traditional forms of media and telecommunication. Second, although everything virtual is ultimately digital in nature (cf. Figure 3-2), many forms of digital technology are not media and the term 'digital' is too broad to be of much value in defining new media, let alone to understand what is so special about it. This is also evidenced by the fact that many entities are digital without thereby being virtual, such as a digital radio or digital camera. Thus, a more promising way of differentiating between old and new media is to look at the fundamental way in which information transfer takes place in the respective forms.

In classical definitions of media, the information exchange is seen as a linear process where a *sender* sends a *message* through a *medium* to a *receiver*. For instance, in the influential transmission model of communication, originally developed by Shannon and Weaver (1949/1998), communication is defined as a message that originates from an *information source,* is encoded by a *transmitter,* sent through a *channel,* decoded by a *receiver* and ultimately arrives at a *destination*. In this model, as well as in daily language, the medium is regarded as a *channel* that we transfer information *through*. This is precisely what is subverted in new media. New media are not channels, but more akin to *places*. Virtual entities can be located in a geometric three-dimensional environment that literally resembles our traditional notion of a *geometric* space. However, they can also be located in a *topological* space, which is an abstract place where virtual entities are *located* and subjected to abstract ordering principles that define their topological location in relation to other virtual entities. For instance, a virtual document can be *within* a virtual folder and *moved* to another, you can go *back* and *forth* between Web sites, a hard drive can be *full* or *empty*, you can *up*load and *down*load files, and so forth (cf. Brey, 1998).

A similar but somewhat misleading view has been presented by Jonathan Steuer, who illustrates the difference between traditional and modern forms of *communication* in Figure 3-1.

As I will return to in more detail below, I think Steuer is mistaken in applying this model to all forms of modern communication and his inclusive use of the term 'virtual reality' blurs rather than clarifies what 'virtual' refers to. A regular conversation on the phone can be described metaphorically as being 'electronically pres-

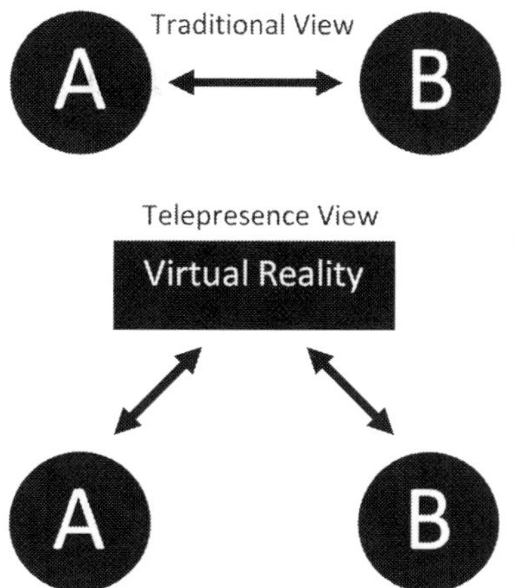

Figure 3-1: The 'traditional' versus the 'telepresence' view of communication (facsimile from Steuer 1992, p. 77)

ent in the same virtual reality' (Steuer, 1992, p. 78), but this metaphorical description is far removed from the way in which virtual worlds *literally* take place in a shared space. However, the model can be used to illustrate one of the crucial characteristics of all things virtual. With virtual environments, in a very literal sense, information is better described as being *made available* rather than *transferred*. As such, the focus in new media, when properly nuanced, is indeed on 'the relationship between an individual who is both a sender and receiver, and on the mediated environment with which he or she interacts' (Steuer, 1992, p. 78). In other words, the medium is not a channel through which we transfer information, but more like a place in which the information remains for a significant duration of time. At the same time, there is no clear distinction between the medium and the information. For instance, if we simply extract the text from a Website or virtual environment, much of the information is lost, including the way it is presented, its topological or geometric relation to other information, and the ways in which we can interact with it. Thus, with virtuality, far beyond a mere metaphorical sense, the message is the medium; the environment constitutes an important part of the information. Strictly speaking, it is not correct to say that these places do not physically exist *anywhere*, since they have to exist in form of a digital representation. However, they do not exist *as meaningful information in* the physical world, let alone as spatio-temporal objects. Indeed, the oft-mentioned oxymoronic features of virtuality arise from the fact that these environments, which constitute *and* contain information, seemingly do not exist in our physical world and can only be accessed using special equipment. This, in turn, is what prompts the familiar spatial metaphors such as cyber*space*, virtual *worlds* and info*sphere*.[2]

Among other things, this model shows how new media can offer many-to-many communication. In many-to-many communication, diverse information cannot be transferred simultaneously from many to many. Instead, a venue must be created in which many can leave information and many can retrieve information; anything else would amount to a cacophony of information overload. This model also conveys how new media can and must be *interactive*, because the same user can be both sender and receiver, and receive available information in accordance with the actions she carries out. Again, manipulating and choosing information (and the way it is presented) in this sense requires that the information is made available rather than transferred through channels. Interactivity entails that you do not choose between 'on' or 'off', which is the primary choice in the earlier sender-receiver model of communication; it entails choosing what, when and how (I will explain this in more detail below).

A review of suggested definitions

Although seeing new media and the virtual as a 'place' rather than a 'channel' brings out some of its important characteristics, this is also in many ways where the seemingly oxymoronic features stem from, and it does not give us a much clearer understanding of what it is or what is so special about it. A number of philosophers and computer scientists have struggled to propose clear and precise definitions, but there is little if any consensus. Before dealing with this problem myself, I will review a selection of promising definitions, and try to extract the criteria that will form the basis of the definitions I will propose towards the end of this paper.[3]

Stanovsky—computer-generated simulation and interactivity

Derek Stanovsky singles out 'computer-generated simulations' and 'interactivity' as the defining terms of virtuality. Note that the qualifier 'computer-generated' has been added to the simulation-criterion, thereby avoiding the criticism I raised against the overinclusiveness of 'simulation'. These, Stanovsky states, *might* be supplemented by three other qualities: being shared by multiple users, providing fully realistic sensory immersion, and enabling users to communicate and act over great distances (Stanovsky, 2004, p. 168). Since Stanovsky explicitly states that the latter are not necessary, he seems to imply that 'computer-generated simulations' and 'interactivity' *are* necessary. Although Stanovsky does not arrive at a concrete definition, his analysis accurately points out what I take to be the two central points of discussion if we are to get a better understanding of what virtuality is. First, in what sense can actual entities be simulated or reproduced in virtual form and what is so special about *computer-generated* as opposed to other forms of simulation? Second, what is so special about the kind of human-computer *interactivity* that is facilitated by virtuality? Without a further elaboration of these points, which I will return to below, one runs the risk of regarding virtual experiences as being 'on more or less equal footing with the more usual forms of experience', or that "all experience is essentially virtual" (Stanovsky, 2004, pp. 172–173).

Brey—three-dimensionality and first-person perspective

Philip Brey defines virtual reality as 'a three-dimensional interactive computer-generated environment that incorporates a first-person perspective' (Brey, 1999, p. 5, 2008) and underlines that full immersion is not to be regarded as a necessary property. Part of the reason for excluding this criterion is that the range of technologies that encompass full immersion is still reserved for very few users. Hence, an ethics dedicated to those systems alone would hardly be worth the effort. He further qualifies his definition by stating that interactivity includes common peripherals such as mouse or gamepad, and that stereo vision, i.e., the actual sensation of being in a 3-D environment, is not a necessary requirement either. Thereby, Brey includes three-dimensional virtual environments that are projected onto a two dimensional computer monitor. Even if projected in this manner, Brey argues that it is essential that the virtual environment itself is three-dimensional, since it fosters an entirely different feeling of immersion and range of life-like actions.

Closely related, it is interesting to note that Brey prefers the concept 'first-person perspective' to 'immersion'. As mentioned above, immersion is a problematic concept since it is primarily a subjective concept—different people become immersed in different things to different degrees. Substituting this concept with 'first person perspective' avoids this problem. The reason why Brey includes the first-person requirement is that it 'suggests a degree of immersion in a world, rather than the experience of a world that can be (partially) controlled from the outside' (Brey, 1999, p. 6). The kind of immersion Brey attributes to having a first-person perspective is clearly important, but the term itself is slightly ambivalent. First, most forms of screen-based VR offer the opportunity to choose freely one's point of view; whether or not one sees the world through the eyes of a virtual body—an *avatar*—does not seem to make much difference. Seeing the world through the eyes of an avatar only makes a profound difference if the perspective replaces our real-world view, e.g., by using a head-mounted display or special VR goggles. However, on this strict interpretation of first-person perspective, which I will later refer to as a first person *view*, many of the applications Brey counts as instances of VR would be excluded. Thus, the reason for using the first-person perspective will prove important in our definition, but due to the ambiguity of the term, I will later suggest the term 'indexicality' instead.

Sherman and Craig

Finally, Sherman and Craig offer the following definition in their *Understanding Virtual Reality* (2003):

> ...a medium composed of interactive computer simulations that sense the participant's position and actions and replace or augment the feedback to one or more senses, giving the feeling of being mentally immersed in the simulation (Sherman & Craig, 2003, p. 13).

Although this definition provides further insight into the essence of virtuality, its meaning differs dramatically depending on how we interpret 'participant'. If, by 'participant's position', Sherman and Craig refer to the position of our physical bodies, then the definition excludes any kind of virtuality in which actions are mediated by an avatar. On this reading, virtuality refers only to those technologies that employ a body suit or other means of motion tracking. Admittedly, this is a distinct and important characteristic of what I will refer to as the subcategory virtual *reality*, but it excludes many of the most interesting and widely available forms of virtuality, in particular any form of virtual environment experienced through a computer monitor. Thus, Sherman and Craig's definition is more apt to define the high-tech forms of virtual *reality*, such as the CAVE,[4] rather than virtuality in general. Sherman and Craig do strike at the core of what virtuality is about in describing it as 'a medium composed of interactive computer simulations' and I will shortly return to their important emphasis on the active role of the computer simulation.

'Virtual' as interactive computer simulation

It is interesting to note that all the definitions discussed include some form of 'computer simulation' and 'interactivity'. I will argue that these are indeed the defining criteria for a generic definition of 'virtual', but, as with the other criteria reviewed, none of these are sufficient on their own. There are of course entities that are interactive without being computer simulations, for instance a car, and entities that are computer simulations without being interactive, for instance when a calculation is performed on the basis of pre-programmed parameters without human intervention. The combination of the two however—interactive computer simulations—will provide the foundation of my definitions and taxonomy of virtuality. Based on previous remarks on the over-inclusiveness of mere 'simulation' and the unique, *interactive* characteristics of virtuality as a new medium, this is a promising starting point for finding a lowest common denominator of all things virtual. 'Simulation' is also not a precise description of virtuality on its own, since it includes a number of situations that cannot readily be described as virtual. In other words, the unique epistemological and ontological status of virtuality disappears if we analyse the phenomenon on a par with, say, an architect's use of miniature models. At the same time, 'simulation' is a more precise term than 'almost the same as' while retaining the notion of something virtual as actual in *some* respects but not in others (as reflected in Lanier's definition above). 'Simulation' accurately captures this property because it entails an analysis of x in order to infer to y. That is, for an x to be a simulation of y, this requires that x cannot be exactly the same as y *and* that x must have some relevant properties in common with y. A miniature airplane will not be a simulation of an actual airplane if the two are the same, nor if the miniature airplane does not have any relevant properties in common with the actual airplane.

To clarify in what sense virtual entities and environments are essentially simulated in nature, a few clarifications are in order before I turn to how additional features constitute different subcategories of virtuality. First, what are they simulations *of*?

Computer simulation of types and tokens

It might sound counterintuitive to define anything virtual as an interactive computer simulation, since this entails that all virtual entities are simulations *of* something. How does this square with the fact that virtual entities do not always correspond to actual entities? After all, we can have computer simulations of all kinds of nonexisting entities, such as fantasy realms and fantasy creatures. This objection is valid, but only if we presuppose a notion of simulation in which the simulated entity must exist as a physical entity. On such a presupposition, a virtual house would fit the definition, whereas a virtual zombie would not. Thus, the notion of 'virtual' as 'simulation of actual' requires some important clarifications. On my usage of the term, 'simulation' can range from token-token to type-type relationships—and, slightly more controversially, it can include simulation of both existing (concrete) and nonexisting (abstract) entities.

First, the distinction between types and tokens is a distinction between a general kind of thing (e.g., horses) and its particular instances ('Clever Hans' or some other particular horse). For instance, we can talk of the horse *type* being thousands of years old despite the fact that no horse *token* would live to that age. There are primarily two competing views on the relation between types and tokens (cf. Wetzel, 2007). On a Platonic view, types (or 'forms') exist independently of their tokens, if any, and do not exist anywhere in space-time. On an Aristotelian view, the types cannot exist independently from their tokens. According to Aristotle, if all tokens of a type (all physically existing horses) were to cease to exist, then the type ('horse') would cease to exist as well. According to Plato, whether the token exists or not does not determine whether the type ('the Idea') exists. What is interesting about these differing views is that from a Platonic perspective, there is no fundamental difference between the type 'horse' and the type 'pegasi', because whether or not they exist is irrelevant. From an Aristotelian perspective, however, this presents a curious dilemma. What kind of concept is 'pegasi' if it has no physical instances? For present purposes, this problem prompts a second distinction between 'abstract' and 'concrete'.

The distinction between abstract and concrete is equally controversial, and a number of different interpretations have been put forward (cf. Rosen, 2006). According to what Lewis calls the 'way of negation' (Lewis, 1986), 'abstract' is typically defined in terms of what it is not—especially which physical properties that are lacking. Perhaps the most promising definition along these lines is that abstract tokens are distinguished by the fact that they have no spatiotemporal location and (hence) no mechanico-causal powers. For instance, a physical horse exists in space-time and can bring about all kinds of changes in the physical world, whereas an abstract Pegasus does not exist in space and can bring about no changes in the physical world. On this definition, 'concrete' refers to any (type of) entity that occupies physical space and is, consequently, capable of causing mechanico-causal effects in the physical world. 'Abstract' refers to all entities that are not concrete in this manner, be they types (e.g., centaurs) or tokens (e.g., the centaur Chiron).

If we follow the distinctions between concrete/abstract and type/token, we get four different kinds of simulation:

> **Concrete type simulations**: Simulation of types that are instantiated in concrete tokens with determinate spatiotemporal location and causal powers (virtual horses, virtual mountains etc.)
>
> **Abstract type simulations**: Simulation of types that are not instantiated in concrete tokens, thus have no determinate spatiotemporal location and causal powers (virtual pegasi, virtual fantasy realms etc.)
>
> **Concrete token simulations**: Simulation of concrete tokens that have a determinate spatiotemporal location and causal powers (virtual 'Clever Hans', virtual Mount Everest etc.)
>
> **Abstract token simulations**: Simulation of abstract tokens that have no determinate spatiotemporal location and causal powers (virtual Pegasus [son of Poseidon], virtual Valhalla etc.)

A fourfold table might clarify further:

	CONCRETE	ABSTRACT
TYPE	Horses, Mountains	Pegasi, Fantasy realms
TOKEN	'Clever Hans', Mount Everest	Pegasus, Valhalla

Table 3-1: Fourfold diagram of types/tokens and concrete/abstract

The important point is that any entity or environment that is virtual is necessarily computer-simulated in one of the senses above. For instance, a virtual library is an interactive concrete *type* simulation (a simulation of something that does exist, but not of one particular entity) whereas the virtual Amsterdam library is an interactive concrete *token* simulation (a simulation of something that does exist, and which is a particular entity, i.e., the actual Amsterdam library). A virtual pegasus is an interactive, abstract *type* simulation (a simulation of something that does not exist and which is not a particular entity) whereas a virtual Pegasus (the token son of Poseidon) is an interactive, abstract *token* simulation (a simulation of something that does not exist, but which is a particular entity). I will later use these distinctions to indicate different forms of virtual-actual relations.

Finally, it could also be objected to my definition that some of the things we refer to as virtual are not in themselves computer-simulated and interactive. For instance, a virtual community is not, strictly speaking, a computer-simulated, interactive community. However, a virtual community is virtual in the sense that it is *made possible by* an interactive computer simulation—in the same sense as an actual community is made possible by the shared venues, shared community objects and so forth. Indeed, one of the essential characteristics of these kinds of social constructed entities, institutional entities in particular, is that their nature is only partly, if at all, determined by their physical properties (Searle, 1995). More generally, in much the same manner that many actual entities and events require an explanation beyond being made possible by physical matter, many virtual entities require an explanation beyond being made possible by an interactive computer simulation.

The computational underpinning

A significant degree of regularity is of fundamental importance to the epistemological status of virtual entities and is secured by the role of the computer that underpins the virtual worlds. This can perhaps best be described by way of a comparison with different philosophical accounts of regularity in the physical world. The similarities and differences between virtual and physical regularity can be illustrated by way of the age-old philosophical discussion of how causality is possible in the physical world. David Hume famously argued that our notion of causality is based on nothing but events that closely succeed each other and therefore become associated into what we refer to as cause and effect:

> Suppose two objects to be presented to us, of which the one is the cause and the other the effect; it is plain, that from the simple consideration of one, or both these objects we never shall perceive the tie by which they are united, or be able certainly to pronounce, that there is a connexion betwixt them (Hume, 1739/2008)

If we see a billiard ball collide with another ball, we cannot experience the supposed causality between them. In other words, there is no magical application of a physical law that determines that certain causes have certain effects. Nicolas Malebranche (1688/1997) struggles with the same problem and finds himself having to postulate an omnipotent God that must actively intervene and make sure that causes are followed by regular effects. Whatever the correct account of causality in the real world might be, Hume and Malebranche point to an important characteristic of virtual worlds. If two billiard balls collide with each other in a virtual world, there *is* a connection between them, namely the computer event that is triggered on that specific occasion and determines what the effect of that cause is going to be. That is, the computer fulfils more or less the same function as Malebranche's God, by constantly having to intervene on every occasion where a particular effect should follow a cause. In more technical terms, virtual environments are characterized by the fact that there is a causal engine that determines the effects of certain causes, for instance by altering the movement of one ball when hit by another on the basis of the mathematical properties of the balls (velocity, direction, impact angle, mass etc.).

This unique characteristic of computer-generated simulation is precisely what makes the virtual fundamentally different from mere products of the mind and at the same time fundamentally different from the real world. The equivalent of Malebranche's God and whatever it is that Hume is missing when we experience an effect following a cause in the physical world can be found within the computer technology that underpins a computer simulation. Although computer-simulated laws of physics are different from laws of physics (e.g., more irregular and subject to lack of accuracy), they can still add a high degree of regularity. That is, the computer simulation can lend stability, persistence, predictability, and intersubjective availability to virtual events and entities, thereby separating the virtual from mere dreams and hallucinations. In more philosophical terms, the computer simulation adds what Kant (1781/1997; KdRV) referred to as congruence between experiences.[5] That is, events do not happen at random, but are results of long, complicated and interconnected chains of events—and these events are subject to certain regularities that allow us to predict and explain the consequences of events and actions.

Most importantly, this entails that facts about virtual worlds can be epistemologically objective. If I claim that my virtual house is located on top of a virtual mountain on a specific island in *Second Life*, this is either true or false for two interrelated reasons. First, anyone can check whether this is the case or not (it is intersubjectively available) and second, the epistemological objectivity of these kinds of claims is made possible by the fact that the computer simulation gives rise to a regularity that allows for both accurate verification/falsification and prediction. This is also the reason why virtual environments are apt for scientific modelling, since the regularities can be created so as to resemble physical laws. Ultimately, it is this regularity that makes virtual acts worth undertaking, since they could not bring about desired consequences if there were no regularity and predictability. Indeed, it is this kind of regularity that makes virtual environments into something more than a spectacle—something more than the irregularity of dreams, illusions and hallucinations. This is in line with Kant's view that certain organizing principles (time, space, causality, permanence etc.) are necessary 'conditions of possibility' to make our experiences intelligible.

Thus, the reason why it is so important to include the computer in our definition of virtuality is to emphasize this point. The 'ultimate reality' of a virtual entity is rooted in the physical world, but only in a form that is not directly accessible (e.g., as strings of binary digits). Thus, truth claims about virtual entities are very similar to claims in general about the physical world. Our claims are usually not about whatever ultimate reality lies behind the world as it appears to us, but they are precisely about how that ultimate reality manifests itself in the world as it appears to us. For instance, it is still somewhat of a mystery what a gravitational force *really* is, but this does not prevent us from making truth claims and predictions on the basis of how the ultimate reality behind gravity manifests itself in the world as it appears to us. Just as we sometimes need instruments to assess the veracity of such claims, we need 'instruments' (screens, networks, peripherals etc.) to assess the veracity of claims about virtual worlds.

In short, it is the computer that underpins the virtual world that facilitates its epistemological objectivity and, in that respect, makes virtual worlds *similar* to the physical world and dramatically *different* from dreams, hallucinations and other products of the mind. At the same time, there are clearly differences between the virtual and the physical, precisely because a computer simulation is a necessary condition for the existence of the virtual world—and because the computer, in contrast with the laws of physics or Malebranche's God, is restricted by the available computational resources.

Interactivity

Although the importance of the computational underpinning captures some of the essence of virtuality, ranging from Websites to immersive VR, it is still limited in the sense that if these technologies were to provide nothing but passive perception, many of the interesting aspects of virtuality would be lost. It is certainly fascinating to watch three-dimensional computer-generated hologram artworks, but as long as they cannot be interacted with, they do not raise the same kind of ontological, epistemological and ethical puzzles. As Dominic McIver Lopes puts it: 'If virtual reality offers anything new it is the possibility for interaction with the occupants and furniture of the computer-generated environment', which is made possible 'precisely because of the special capabilities of computing technology' (Lopes, 2004, p. 110).

Most obviously, the ability to interact with the computer allows humans to communicate and interact with each other over great distances *through* the computer. Thus, what I will refer to as a 'virtual world' differs enormously from a hologram because of the ability to share the experience with others from across the globe. Although 'networked communication' is not necessary for something to be virtual, as I will return to below, it is an important part of virtual worlds and virtual communities, and *interactivity* is necessary for networked communication. Another obvious advantage of interactivity is that it is required in order to have an immersive three-dimensional, or what I will refer to as an indexical, experience. Simply put, you cannot 'look around' if the computer simulation has no idea where you or your virtual representation (the avatar) is looking, which requires interaction—either indirectly by some kind of steering mechanism or directly by means of motion detectors that report the direction of your head/eyes to the computer simulation. More fundamentally, some level of interactivity is necessary for all kinds of virtuality. In computer science terminology, an interactive computer simulation is sometimes aptly described as a computer simulation with a 'human-in-the-loop'.[6] Without a human in the loop, a computer simulation (running on a finite state computer) will be entirely deterministic in the sense that it will only do whatever it has been programmed to do—determined solely by pre-programmed variables. If the same variables are used, the computer simulation (running on a deterministic computer) will produce exactly the same result on subsequent runs.

Jonathan Steuer defines interactivity in terms of 'the extent to which users can participate in modifying the form and content of a mediated environment in real time' (Steuer, 1992, p. 84).[7] Defining human-computer interaction as the ability to modify the form and content of a mediated environment can be seen as setting the threshold too high, since a number of technologies that are sometimes referred to as virtual do not meet the requirements. Although I do in general strive to define concepts according to common usage, this is a case where the term virtuality *ought* to be restricted. Some of the technologies hailed as interactive are in fact not. For instance, most forms of streamed media on the Internet are, in all important respects, the same as traditional forms of media if the user cannot, in a non-trivial manner, choose what to watch, when to watch it, and in which form to watch it—to, as it were, seek out the place where the content exists and interact with it. If we regard interactivity as a necessary requirement for virtuality, it follows that online, streaming, live media, for instance, do not qualify. Thus, terms like 'virtual TV channel' or 'virtual radio' are misnomers if the only thing that separates them from traditional media is that they are transmitted in digital rather than analogue form. The proper terms would be 'digital TV channel' or 'digital radio'. If we were to include these media in our definition of virtual, we would be left with a definition along the lines of 'not physical', or worse, 'something other than the old' since a 'real' TV channel is not physical either. Indeed, I believe the way in which 'virtual' has become a prefix for any kind of digitized version of traditional media and other phenomena is part of the reason for its present vagueness and ambiguity.

That being said, 'interactivity' should not be defined too strictly. In what I will shortly define as virtual environments (which includes, e.g., three-dimensional computer games), it is important that the environment is what Dominic McIver Lopes characterizes as *strongly* interactive. Strong interactivity entails that the 'users' inputs help determine the subsequent state of play' (Lopes, 2001, p. 68). In short, the state of affairs in a virtual environment should be largely determined by the users' actions. At the very least, this pertains to the location, orientation, abilities and so forth of your avatar, but the interactivity will of course be particularly strong when the state of the environment is fundamentally molded by the actions of the user. This is primarily the case in what I will refer to as persistent, multi-user virtual *worlds* where the state of the environment is a result of many years of user input from multiple users. When it comes to virtual *entities,* we should allow for a somewhat weaker sense of interactivity. This can be illustrated by way of game researcher Espen Aarseth's distinction between fictional and simulated objects (Aarseth, 2006). In the classic game *Return to Castle Wolfenstein,* the doors in the game are sometimes interactive (they can be opened, closed and so forth) but other doors are simply textures on the wall—and not interactive in a strong sense of the word. However, these doors may still be referred to as being computer-simulated and interactive, in the sense that they will stop your avatar from walking through them. The interactivity lies in the fact that the computer simulation must actively prevent your avatar from walking through the door. Hence, there is an interactive process between your avatar-mediated actions and the computer simulation that underpins the virtual environment.

'Virtual' defined

To summarize so far, we can define 'virtuality', and the adjective 'virtual' as interactive computer simulations. The definition consists of three elements:

- **'Computer':** The ontologically objective, physical 'grounding' that facilitates regularity and intersubjective availability, hence epistemological objectivity.
- **'Simulation':** entails that virtual entities are similar to actual entities in some respects and different in other (partly determined by whether they are abstract or concrete, types or tokens).
- **'Interactivity':** facilitates networked computing, immersion (indexicality) and virtual acts as modification of the form and content of a virtual environment.

Virtuality in this generic sense raises many philosophical problems on its own, primarily theoretical ones. However, the generic definition hides the fact that virtuality comes in many different forms, each presenting unique philosophical issues. In what follows, I will outline what I regard to be the three most important kinds of virtuality: virtual *environments,* virtual *worlds,* and virtual *reality*—which correspond to the criteria 'indexicality', 'multi-access' and 'first-person view'.

Virtual environments, indexicality and avatars

Recall that Brey singled out three-dimensionality and having a first-person perspective as a requirement for virtual reality, due to their ability to foster a form of immersion in the virtual environment. Although I agree with the reason for including these criteria, and find them more precise than the more commonly used 'immersion', the notion 'first-person perspective' is slightly misleading if defined loosely and too exclusive if defined strictly. First, if we define first-person perspective strictly (which I will later refer to as first-person *view),* it requires one to have the ability to literally look around a virtual environment and change perspective by moving your head and/or eyes as you would in the physical world. Such a conception would exclude many of the applications that Brey includes, in particular all non-immersive (screen-based) forms of VR. If we define the term more loosely, however, the term becomes slightly misleading. Whether or not you *perceive* yourself as acting from a first- or third-person perspective in a screen-based form of VR does not make much difference. What makes a difference lies in the fact that a virtual environment can be 'interacted with from a single locus' (Brey, 1999, p. 6).

Interacting from a single location in the virtual environment without having a first-person *view* requires some kind of virtual entity that you control and through which you interact with other virtual entities. This is known as an avatar, which is the representation of the human user in a virtual environment, either in the form of a graphical object (usually in human-like form) or in the form of a nickname in non-graphical environments such as discussion forums and chat rooms.[8] In virtual worlds such as *Second Life* all interactions are carried out as if done from the location of your avatar, but you are free to choose whether you perceive the world from your avatar's eyes, from behind your avatar's shoulders, or from any 'God's eye' point of view. What is important is that the agent interacting with the environment is itself located in the environment.[9] Using an avatar that is located in a three-dimensional environment is important because it means that 1) the movements and actions of your avatar can be restricted in many of the same ways as in real life; 2) your avatar can be interacted with by the environment and, if in a multi-access environment (see below), seen and interacted with by others; 3) you can often configure your avatar to look however you wish thereby possibly creating stronger emotional ties between yourself and your representation; and 4) your avatar can engage in bodily acts that are similar to real life actions. These points, especially the latter, are responsible for many of the ethical problems that can arise in virtual environments, as for instance when users commit acts with their avatar that are unethical and/or illegal in the real world, such as virtual violence and virtual paedophilia.

Due to the ambiguity of terms like 'immersion' and 'first-person perspective', and the important differences caused by being an agent *in* a virtual environment, I propose 'indexical' as an alternative criterion. Admittedly, the term does not make immediate sense, but I have chosen it for two reasons. First, the term does not suffer from the ambiguity caused by the metaphorical use of terms like 'immersion' and 'first-person perspective' and I will reserve the notion of first-person *view*, defined in a strictly non-metaphorical manner, as the criterion for fully immersive virtual environments only—which I will refer to as virtual *reality* (see below).

Second, philosophers such as Ernst Tugendhat (1976) and Truls Wyller (2000) have given a precise meaning to the term that perfectly captures the importance of acting from a single locus. In philosophy, 'indexicality' is used in two different but related ways. In philosophy of language (and linguistics), indexicality refers to words whose meaning depends on the context in which they are uttered. For instance, saying that there is a tree to my left does not provide any knowledge about the location of the tree unless you know my spatio-temporal location and orientation. This means that terms such as 'there', 'here', 'to my left' and 'above me' do not make literal sense unless the person who uses the words has a known location and orientation in a three-dimensional space. This is what has spurred a second use of the term 'indexicality', inspired by neo-Kantian philosophers such as Tugendhat and Wyller. In this second use, indexicality is not only a property of words, but also a property of our relation to our surroundings. As an example, trying to orient oneself by use of a map is entirely pointless if one does not know one's location on the map. Thus, stationary maps typically have a 'you are here' marker, which serves as an *index* from which you can orient yourself. In this sense of the word, 'indexicality' means to have a discreet, subjective (or egocentric) position from which we act and orient ourselves in a three-dimensional world (cf. Wyller, 2000, p. 39). In this form, the notion 'indexicality' captures both the three-dimensionality and the first-person criteria, and does not carry the metaphorical connotations implicit in terms like 'immersion'. When coupled with the interactivity criterion, which is a necessary criterion for all things virtual, indexicality also means that one is an *agent* at a specific place and, as such, *is present* at that place in one form or another—typically as an avatar. Thus, a first subcategory of virtuality is that which is not only computer-simulated and interactive, but also *indexical* as defined above. Since indexicality requires a three-dimensional space, I will refer to such an interactive and indexical computer simulation as a *virtual environment*.[10]

The reason I do not include indexicality as a criterion for virtuality in general is that terms such as virtual libraries, virtual banks and virtual universities make perfect sense even if they are not three-dimensional. Indeed, in virtue of being what Brey (2003) refers to as 'ontological reproductions', these are among the most important phenomena when it comes to the impact of virtual entities on our daily lives. These are the types of entities that have been socially constructed (a library, bank or university would not be what they are if there were no conscious beings), which means that their existence is not closely tied to particular physical properties. The same type of social reality can also be constructed from within what I will refer to as virtual *worlds*, as made possible by the presence of *multiple users* within a virtual environment.

Virtual worlds, multi-access and virtual communities

Virtual environments, as defined above, have been prevalent since the 1980s, in the form of single-player computer games. The addition of multiple access in virtual environments only became widely popular with the introduction of affordable broadband in the late 1990s. Text-based virtual communities, however, have been relatively widespread since the 1980's—famous examples including the WELL (Whole Earth 'Lectronic Link) and LambdaMOO. By 'multiple access' (or the 'multi-access-criterion') I mean the simultaneous presence of multiple participants in a computer-simulated interactive environment (three-dimensional or otherwise), where the participants can communicate with each other. In an interactive computer simulation, this requires (graphical) representations of the participants and the use of networked communication.[11] It also requires (or, facilitates) a 'persistent space', meaning that you normally cannot *pause* a virtual world. Indeed, the reason I have chosen 'multiple access', rather than 'multiple users' or similar, is that access captures the importance of there being a persistent world that is persistently 'there' for users to access—a world embodying objects that continue to exist, and events that transpire also when you are logged out. Virtual worlds exist for users to go in and out of, and your actions and their consequences take place in the same space as the actions and consequences of other users. Thus, multi-access, and hence persistent space, makes the place metaphor even more apt; the 'place' exists whether you are there or not, and not only in a topological meaning. This also explains why virtual worlds have to be real-time, since they must allow one user's actions to be immediately visible to other users. The multi-access criterion is difficult because it cuts across very differing forms of virtuality, and is absent even in some of the most technologically advanced forms. There is a profound difference between multiple users that, on the one hand, meet and communicate in a virtual world proper, and on the other, multiple users that meet and communicate in a non-graphical discussion forum or chat room. What they have in common is that users communicate through a representation of themselves, thus allowing for pseudonymous communication. However, the philosophical and ethical challenges become much more complicated with the introduction of a *graphical* representation, since this allows for re-creation of acts, events and experiences that cannot be carried out in the actual world, be it for physical, economical or ethical reasons. Hence, I will use the established term 'virtual community' to describe communication between multiple users that is computer mediated and interactive, but not necessarily taking place in a three-dimensional environment. Because of the significant difference between virtual communities on the Web and virtual communities inside virtual *environments,* I propose that computer-simulated, interactive, indexical, *multi-user* environments should be referred to as *virtual worlds.*[12] Among the reasons for choosing 'world', in contrast with environment, is that 'world' is often used to signify a collection of people, or the earth *as inhabited*—hence expressions like 'the modern world' and 'the ancient world', which implies different worlds on the same earth.[13]

I have defined virtual worlds as interactive, computer-simulated, indexical, multi-access environments, which means that all virtual worlds are graphical, three-dimensional, persistent environments that allow for user-to-user interaction. Although these features are shared by all virtual worlds, it is important to note that there are still a number of differences between the many hundred different virtual worlds out there.[14] Indeed, Sarah Robbins-Bell (2008) is developing a faceted classification scheme of virtual worlds, according to which virtual worlds are classified according to such categories as physics, type of objectives, dominant form of communication, object ownership, avatar customizability, access model, user-to-user relationship and formation of communities. Although *Second Life* and *World of Warcraft* are both virtual worlds according to my definition, I agree with Robbins-Bell's elaboration of the nuances within this category, according to which *Second Life* and *World of Warcraft* differ in almost all respects.

So far, I have defined 'virtual' as something that is computer-simulated and interactive, and 'virtual environment' as a subcategory that requires indexicality in addition. 'Virtual worlds' are, in turn, a subcategory of 'virtual environments', further requiring multiple access as outlined above. I now turn to 'virtual reality', which is also a sub-category of 'virtual *environment*' in that it requires indexicality. However, 'virtual reality' does not require multi-access, but instead what I will refer to as a genuine 'first-person view'.

Virtual reality and first-person view

For virtual reality, indexicality is necessary but not sufficient; it requires your indexical location, orientation and movements in the virtual environment to actually correspond to the indexical location, orientation and movement of your physical body—it requires what I will refer to as a genuine 'first-person view'.

Having a first-person view, in a strict sense, means that what you perceive is determined by the location and orientation of your eyes. For example, a first-person view restricts you from seeing an object from different angles without either rotating the object or moving your eyes and body. This strict definition entails that, in order to have a first-person view in a virtual environment, 1) you must be in a three-dimensional environment, 2) you must physically move (a part of) your body in order to look around and navigate, and 3) your movements and location in the virtual environment corresponds to your movements and location in the actual world. That is, there is a 'concrete token' relationship between your movements and location in the virtual reality and the movements and location of (parts of) your physical body. Thereby, 'first-person view' encompasses all the essential characteristics of high-end forms of virtual reality and renders indexicality as redundant for the definition of virtual reality. This is reflected in the fact that all high-end virtual reality systems require a device that directly feeds the virtual environment to your eyes, by means of either a head-mounted display or VR goggles.[15] When using transparent goggles that convert a stereographic two-dimensional image into the appearance of a three-dimensional environment, you move and look around just as you would in real life. Head-mounted displays, however, require motion tracking in order to adjust the participant's field of view according to the movements and orientation of the head. Having a first-person view in this sense truly makes the virtual experience lifelike. This is reflected in the fact that virtual reality first-person view can induce, and is often used to treat, phobias such as fear of heights or arachnophobia (see e.g., North, North, & Coble, 1998)

Based on the considerations above, I take first-person view to be the minimal requirement of virtual *reality*, since it truly *immerses* a participant in the virtual reality, far beyond any metaphorical sense. The importance and necessity of the connection between movement of the eyes and spontaneously altered perception is nicely captured by Edmund Husserl:

> If the eye turns in a certain way, then so does the 'image'; if it turns differently in some definite fashion, then so does the image alter differently, in correspondence. We constantly find here this two-fold articulation... Perception is without exception a *unitary accomplishment* which arises essentially out of the playing together of two *correlatively related functions*... In virtue of [these processes] one and the same external world is present to me. (Husserl, 1989, p. 63 [58], italics in original)

An interesting way of illustrating Husserl's phrase 'one and the same external world' is that with the term virtual *reality*, in contrast with virtual environments and virtual worlds, we intuitively resist its plural form. The reason is precisely that being in a reality requires having a first-person view, and it is impossible to have more than one first-person view. That is, you cannot participate in multiple virtual realities simultaneously any more than you can be in more than one spatiotemporal place simultaneously in physical reality. You can have as many avatars in as many virtual worlds or environments as you like, but as soon as you experience the virtual reality 'through your own eyes' you can only be in one virtual reality at a time.

The virtual reality can be augmented by the addition of more comprehensive forms of motion tracking and sensory feedback. By using highly advanced data gloves, one can have tactile perception, for instance by feeling resistance when one grabs an object.[16] It should be noted that motion tracking comes in widely differing degrees. The most advanced forms of VR can implement so-called 'full body immersion' by measuring every relevant movement of the entire body, as well as provide tactile feedback to different parts of the body. The least advanced forms measure only the acceleration of one or very few parts of the body. Indeed, such limited forms of motion tracking are currently the only affordable VR technologies for most ordinary consumers. Off-the-shelf head-mounted displays, if they were to function optimally, would foster a presence in the virtual environment that is qualitatively different from the presence one experiences when the three-dimensional environment is projected onto a monitor and one changes perspective with the use of mouse, keyboard and/or gamepad.[17] However, the use of such head-mounted displays in virtual environments, and especially virtual *worlds,* has many flaws, including time delay ('lag'), artificially narrow field of vision, reduced image resolution due to bandwidth limitations, and reports of considerable eye strain and nausea (Patterson, Winterbottom, & Pierce, 2006; Regan, 1995). If these flaws are sorted out, it is probable that this kind of augmentation of virtual worlds such as *Second Life* will become the most likely point of convergence between virtual worlds and virtual reality in the near future. The addition of multiple access to virtual reality, although still not feasible, clearly makes an important difference for the same reasons that virtual *worlds* are different from mere virtual *environments*. Having multiple access in VR would require body-like representations of all the participants, which would lead to simulation of even more lifelike activities, for instance as made possible by computer mediation of touch. However, since this technology is presently not feasible—at least not beyond a very limited degree—it hardly warrants a term of its own. Insofar as I will refer to such (future) technologies, I will simply use the qualifier *multi-access* virtual reality.

Definitions

After having reviewed some influential definitions of virtuality and analysed and re-phrased the distinguishing characteristics of the criteria found relevant—computer simulation, interactivity, indexicality, multiple access and first-person view—we get the following definitions:

> **Virtual x**: Interactive, computer-simulated x (or, x made possible by interactive computer simulation).[18]
>
> **Virtual environment**: Interactive, computer-simulated, *indexical* environment
>
> **Virtual world**: Interactive, computer-simulated, indexical, *multi-user* environment
>
> **Virtual reality**: Interactive, computer-simulated environment experienced from a *first-person view*.

The relation between these concepts can be further illustrated through a Venn diagram (see Figure 3-2).

In order to illustrate these definitions further, we can arrange a table of the relevant criteria as applied to some of the more popular forms of virtuality, including some non-virtual entities for comparison (Table 3-2).

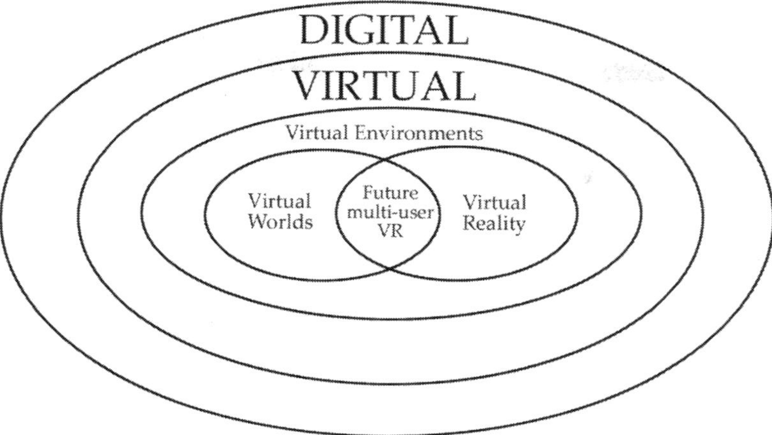

Figure 3-2: Illustration of the relation and intersections between the different categories

CATEGORY	COMPUTER SIMULATION	INTER-ACTIVE	INDEX-ICAL	MULTI-ACCESS	FIRST-PERSON VIEW	EXAMPLES
Multi-access VR	X	X	X	X	X	Future VR (including HDM-augmented virtual worlds when feasible)
Virtual Reality	X	X	X		X	CAVE, Virtusphere, Hi-tech flight simulators
Virtual World	X	X	X	X		*Second Life*, *World of Warcraft*, Sims Online
Virtual Envi-ronment	X	X	X			Most modern, off-line video games, incl. Nin-tendo Wii
Virtual Com-munity	X	X		X		LambdaMoo, chat rooms, discussion forums, social networking sites
Virtual (generic)	X	X				Virtual banks, libraries, uni-versities
Real-time streaming media	X					Digital TV channels
Stereoscopy	(X)		X	X	X	3D movies, holograms
Physical reality		X	X	X	X	
Dreams and il-lusions		X	X		X	

Table 3-2: Taxonomy of 'virtuality' and its subcategories

This table requires some further elaboration. The combination of first-person view and multiple access (to such an extent that it requires networked communication) is currently not feasible due to technological limitations. Thus, multi-access is not a necessary criterion for virtual reality, which thereby includes high-end technologies such as the CAVE and Flight simulators. As explained above, virtual worlds and virtual environments differ from virtual reality in that they do not require a first-person view, and virtual worlds differ from virtual environments in that they allow multiple access. Furthermore, virtual communities differ from virtual environments in that they do not require indexicality, but they are qualitatively different from virtuality in general, which does not require interaction between multiple users. Finally, the 'virtual' category, which encompasses all the other subcategories, requires computer simulation and interactivity, which *together* are necessary and sufficient criteria. That is, an entity is virtual if and only if it requires an interactive computer simulation for its existence. 'Required for its existence' corresponds to how for instance 'social entities' require a social collective for their existence.

I have also included a number of non-virtual entities for comparison. As mentioned, there is no qualitative difference between streaming media and traditional media, since none of them involve interactivity (at least not to any interesting degree). Three-dimensional movies and the like are three-dimensional, can be shared with multiple users and are seen from a first-person view, but they still lack interactivity (hence only limited indexicality) and consequently do not differ substantially from other forms of visual illusions. In order to illustrate how the virtual occupies a middle ground between dream and reality, as elaborated above, the table shows that reality is not a computer simulation,[19] and that dreams and hallucinations lack the consistency, coherence and intersubjective availability supplied by both physical laws and computer simulations, and the presence of multiple users. The important point is that dreams and hallucinations do not allow for 'multiple access', in contrast with both reality and virtual worlds.

It might seem odd that in the figure above, physical reality has *fewer* properties than a multi- access VR, but this merely entails that if all of physical reality were to be simulated in a virtual reality, the *addition of* computer simulation will make the virtual reality different from physical reality—i.e., the differences between the two would be a result of the fact that the interactivity, indexicality, multi-access and first-person views only exist as a result of the computer simulation. In other words, claiming that virtual reality *lacks* something is entirely consistent with the figure above—the lacks might be a result of the addition of the underpinning computer simulation.

Why are these distinctions important?

Although precision and clarity is important in general, it is particularly important when we make claims about the ethical and societal implications of technology—even more so when we make claims about 'virtuality', because such claims often involve a comparison between virtual and actual. This relates to the so-called *principle of formal equality*,[20] which states that a difference in treatment or value between two domains can only be justified on the basis of a relevant and significant difference between the two. For instance, many issues in animal ethics can be approached by first discussing the theoretical differences between humans and other animals (e.g., that some animals have a significantly less developed Central Nervous System) and subsequently discuss to what degree these differences are relevant to their value (e.g., that the less developed CNS indicates little or no ability to experience pain). More generally, for something to be of more or less value than something else, the two must be significantly different in one way or another, and we must be able to justify why this difference entails a difference in value. This entails that when we ask how trust in virtual worlds may differ from trust in actual reality, we are asking 1) how do virtual worlds differ from actual reality, and 2) how do these differences determine the possibility for trust?

A few examples may illustrate this point. Howard Rheingold argues that the lack of spontaneous bodily gestures and facial expression is the reason for the 'ontological untrustworthiness' of virtual acts of communication (Rheingold, 2000, p. 177). This critique seems to hold for virtual communities where there is no bodily representation, but it may not hold for virtual *reality* where motion tracking can relay some spontaneous bodily gestures and even facial expressions. Similarly, Lucas Introna argues that 'virtual communities are... different to those that are situated, embodied and collocated in that they have much less resources available to express and secure their identity through shared community objects' (Introna, 2007, p. 170). Again, this seems to hold for many types of virtual communities, but perhaps not for virtual *worlds*, since the multi-user requirement entails that these virtual worlds have a completely different level of permanence, which may in turn entail the possibility of having relatively stable virtual entities that *can* act as shared community objects. It is not my purpose to discuss the validity of these claims here, but to point out that the relevant considerations will necessarily be different depending on what type of virtuality we are talking about—and it has been the purpose of this paper to contribute a set of definitions and distinctions that can be useful in this regard.

Conclusion

The purpose of this paper has been to propose definitions for the terms 'virtual', 'virtual environment', 'virtual world', and 'virtual reality' as well as to clarify how they relate to each other. I have also argued in more detail how important it is to emphasize the role of the interactive computer simulation as provider of regularity, intersubjective availability (in the case of virtual communities and worlds), and hence an epistemological objectivity that in many ways is equivalent to that of the actual world. Although this may not be the final word on how these concepts ought to be defined, I hope to have shown the importance of providing such definitions, the complexities involved in doing so, as well given a glimpse into the fundamental philosophical challenges that are raised by the characteristics of different forms of virtuality. Clarity in these issues is of utmost importance when we move on to make claims about the epistemological and ontological status of virtual entities, such as when discussing the nature and possibility of 'trust' in virtual environments, virtual worlds or virtual reality.

Acknowledgments

I would like to thank Charles Ess for carefully reading and commenting on this paper, as well as Philip Brey and Adam Briggle for giving feedback on earlier versions. This paper has been presented at the workshop 'Philosophy of Virtuality: Deliberation, Trust, Offence and Virtues' (NTNU, 2009) and the 'Philosophy of Computer Games Conference' (UiO, 2009) and I am grateful for the helpful feedback from both audiences.

Notes

1. See Marianne Richter (this volume) for an overview and discussion of ways in which 'actuality', 'reality' and related terms have figured in the history of philosophy.
2. The term 'cyberspace' was coined by William Gibson in his famous *Neuromancer* (1984). The term 'infosphere' appears to have been coined by R.Z. Sheppard (Sheppard, 1971) and plays an important role in Luciano Floridi's philosophy and ethics of information (see e.g., Floridi, 2002).
3. Cf. Søraker (2010) for a more thorough review and critique of a number of other suggested definitions. I single out Stanovsky, Brey and Burdea & Coiffet here because the discussion of their definitions will be directly relevant for my proposed definitions later.
4. The CAVE is an immersive virtual reality where the images are projected on the walls of a cube surrounding the user. Such systems usually do not employ networked communication due to the immense speed and bandwidth that would be required, which also illustrates why

networked communication is not an apt criterion for 'virtual reality'. Cf. http://www.evl.uic.edu/pape/CAVE/ [Retrieved May 22, 2010].

5. The notion of congruence recurs throughout KdRV, but see in particular B278-279 ('congruence with the criteria of all actual experience'), A376, A112 and A451/B479. In the words of Kant, the 'connection of appearances determining one another with necessity according to universal laws [is] the criterion of empirical truth, whereby experience is distinguished from dreaming' (KdRV, A493/B521).
6. Cf. DoD Modeling and Simulation Glossary (United States Department of Defense, 1998, p. 124).
7. The importance of real-time simulation will become clearer in defining virtual worlds as essentially 'multiple access' below.
8. The term 'avatar' as referring to virtual agents was popularised by Neal Stephenson in his influential novel *Snow Crash* (1993). The term originally comes from Hindu and refers to the physical incarnation of a divine being.
9. As Espen Aarseth has reminded me, in some computer games you interact with the virtual environment from a God's-eye point of view and/or control a collection of 'avatars', e.g. in real-time strategy games such as Command & Conquer. Although these environments lack indexicality as defined above, they do embody three-dimensionality and many of the other features of virtual environments, thus I suggest referring to these simply as non-indexical virtual environments or, if massively multiplayer, as non-indexical virtual worlds.
10. There are, or rather were, non-graphical, one-player environments (e.g., text-based adventure games or 'interactive fiction', such as *Galatea*) that embody many of the same properties as virtual environments, including a weak form of indexicality. These are relatively rare, and increasingly so, hence do not require a term of their own and can be referred to as *non-graphical* virtual environment. If such non-graphical environments allow for multiple users (e.g., online text-based role-playing games such as LambdaMOO), they will be covered by the 'virtual community' category as explained below.
11. Another form of involving multiple users is sometimes referred to as 'hot seat'. This means that a number of participants can participate in the same environment without the use of networked communication. This is, for natural reasons, limited (usually no more than four participants at the same time) since the view will be the same for all participants. Hence, I will not include these technologies under the heading of virtual worlds.
12. I will return to the possibility of multiple users in fully immersive virtual reality below.
13. Indeed, 'world' is made up of the roots of *wer* and *ald*, originally from German, meaning 'man' and 'age'. Thus, the old English 'world' literally meant 'age of man'.
14. See the Association of Virtual Worlds' *Blue Book* (Association of Virtual Worlds, 2008) for a comprehensive list of the many hundred different virtual worlds currently in use.
15. In the distant future, it is not impossible that we will see brain-computer interfaces that feed the visual stimuli directly to the brain instead, as illustrated in the Matrix movies. Indeed, there have been some early attempts to control *Second Life* by means of a brain-computer interface (Keio University, 2008).
16. In continuation of the quote above, Husserl adds that 'the like holds, obviously, for touch' (ibid).
17. There are also other forms of motion-tracking that foster some degree of immersion, and as such constitutes a grey area between virtual environments and virtual reality. To take but one

example, Microsoft's upcoming release of 'Project Natal', an addition to the XBOX console, promises interaction solely through bodily gestures and speech recognition. These kinds of technologies allow for a concrete type relationship between your body and your avatar, since the avatar's action will typically correspond only to the type, not the token, activity of your body. Currently, this does not count as virtual reality proper, however, since it lacks a pure first-person *view* as described.

18. This definition may or may not exclude some entities commonly referred to as 'virtual'. It intentionally excludes things like 'virtual radio', which is a misnomer (see above). It also excludes terms like 'virtual memory', 'virtual server' and other computer science terms. These are established terms and I simply grant that they use the term 'virtual' in a different sense from the way I have defined it above. 'Virtual memory' and the like seem to rest on the common sense definition of 'almost the same as', or 'fulfilling the same role as' actual computer memory, which can be very misleading when talking about virtual worlds and entities therein, as discussed above.
19. Although it is logically possible that reality is computer-simulated, as argued by Nick Bostrom (2003), this is not an option I will entertain in this paper.
20. The principle is usually attributed to Aristotle (*Nicomachian Ethics,* V.3. 1131a10-b15; *Politics,* III.9.1280 a8-15, III. 12. 1282b18-23; cf. Gosepath (2008)). See also Søraker (2007) and Wetlesen (1999).

References

Aarseth, E. (2006). Doors and Perception: Fiction vs. Simulation in Games [Electronic Version]. *Digital Arts & Culture Conference*. Retrieved September 2, 2009 from http://www.luisfilipeteixeira.com/fileManager/file/fiction_Aarseth_jan2006.pdf.

Association of Virtual Worlds. (2008). *The Blue Book: A Consumer Guide to Virtual Worlds*. Retrieved October 25, 2008, from http://associationofvirtualworlds.com/thebluebook/the_blue_book_august_2008_edition.pdf.

Borgmann, A. (1999). *Holding On to Reality: The Nature of Information at the Turn of the Millennium*. Chicago, IL: University of Chicago Press.

Bostrom, N. (2003). Are You Living in a Computer Simulation? *Philosophical Quarterly, 53*(211), 243–255.

Brey, P. (1998). Space-Shaping Technologies and the Geographical Disembedding of Place. In A. Light & J. M. Smith (Eds.), *Philosophy and Geography III: Philosophies of Place* (pp. 239–263). Lanham, MD: Rowman & Littlefield.

Brey, P. (1999). The Ethics of Representation and Action in Virtual Reality. *Ethics and Information Technology, 1*(1), 5–14.

Brey, P. (2003). The Social Ontology of Virtual Environments. *American Journal of Economics and Sociology, 62*(1), 269–282.

Brey, P. (2008). Virtual Reality and Computer Simulation. In K. E. Himma & H. T. Tavani (Eds.), *The Handbook of Information and Computer Ethics* (pp. 361–384). Hoboken, NJ: John Wiley & Sons.

Floridi, L. (2002). On the Intrinsic Value of Information Objects and the Infosphere. *Ethics and Information Technology, 4*(4), 287–304.

Gibson, W. (1984). *Neuromancer*. New York: Ace Book.

Gosepath, S. (2008). Equality. In E. N. Zalta (Ed.), *The Stanford Encyclopedia of Philosophy (Fall 2008 Edition)*.

Hume, D. (1739/2008). *A Treatise of Human Nature*. Sioux Falls, SD: NuVision.

Husserl, E. (1989). *Ideas Pertaining to a Pure Phenomenology and to a Phenomenological Philosophy [Second book]* (R. Rojcewicz & A. Schuwer, Trans.). Dordrecht, Netherlands: Kluwer.

Introna, L. (2007). Reconsidering Community and the Stranger in the Age of Virtuality. *Society and Business Review, 2*(2), 166–178.

Kant, I. (1781/1997). *Critique of Pure Reason* (P. Guyer & A. W. Wood, Trans.). Cambridge: Cambridge University Press.

Keio University. (2008). Press Release: Keio University Succeeds in the World's First Demonstration Experiment with the Help of a Disabled Person to Use Brainwave to Chat and Stroll through the Virtual World [Electronic Version]. Retrieved August 1, 2008 from http://www.keio.ac.jp/english/press_release/080605e.pdf.

Lanier, J. (1999). Virtual Reality. *Whole Earth* Fall 1999. Retrieved January 6, 2008, from http://findarticles.com/p/articles/mi_m0GER/is_1999_Fall/ai_56457593

Lewis, D. (1986). *On the Plurality of Worlds*. Oxford: Basil Blackwell.

Lopes, D.M. (2001). The Ontology of Interactive Art. *Journal of Aesthetic Education, 35*(4), 65–81.

Lopes, D.M. (2004). Digital Art. In L. Floridi (Ed.), *Blackwell Guide to the Philosophy of Computing and Information*. Oxford: Blackwell.

Malebranche, N. (1688/1997). *Dialogues on Metaphysics and on Religion* (N. Jolley & D. Scott, Trans.). Cambridge: Cambridge University Press.

North, M.M., North, S.M., & Coble, J.R. (1998). Virtual Reality Therapy: An Effective Treatment for Phobias. *Stud Health Technol Inform*(58), 112–119.

Patterson, R., Winterbottom, M.D., & Pierce, B.J. (2006). Perceptual Issues in the Use of Head-Mounted Visual Displays. *Human Factors: The Journal of the Human Factors and Ergonomics Society, 48*, 555–573.

Regan, C. (1995). An Investigation into Nausea and Other Side-Effects of Head-Coupled Immersive Virtual Reality. *Virtual Reality, 1*(1), 17–31.

Rheingold, H. (2000). *The Virtual Community: Homesteading on the Electronic Frontier*. Cambridge, MA: MIT Press.

Robbins-Bell, S. (2008). *Using a Faceted Classification Scheme to Predict the Future of Virtual Worlds*. Paper presented at the Internet Research 9.0 (AoIR) conference, October 15–18, 2008. Copenhagen, Denmark.

Rosen, G. (2006). Abstract Objects. In E. N. Zalta (Ed.), *The Stanford Encyclopedia of Philosophy (Spring 2006 Edition)*.

Searle, J. (1995). *The construction of social reality*. New York: Free Press.

Shannon, C.E., & Weaver, W. (1949/1998). *The Mathematical Theory of Communication*. Urbana: University of Illinois Press.

Sheppard, R.Z. (1971). Rock Candy [Electronic Version]. *Time*. Retrieved July 12, 2009 from http://www.time.com/time/magazine/article/0,9171,905004,00.html.

Sherman, W.R., & Craig, A.B. (2003). *Understanding Virtual Reality*. San Francisco: Elsevier.

Shields, R. (2003). *The Virtual*. London: Routledge.

Søraker, J.H. (2007). The Moral Status of Information and Information Technologies—a Relational Theory of Moral Status. In S. Hongladarom & C. Ess (Eds.), *Information Technology Ethics: Cultural Perspectives* (pp. 1–19). Hershey, PA: Idea Group Publishing.

Søraker, J.H. (2010). *The Value of Virtual Worlds and Entities [Dissertation]*. Enschede: University of Twente.

Stanovsky, D. (2004). Virtual Reality. In L. Floridi (Ed.), *The Blackwell Guide to the Philosophy of Computing and Information* (pp. 167–177). London: Blackwell.

Stephenson, N. (1993). *Snow Crash*. New York: Bantam Books.

Steuer, J. (1992). Defining Virtual Reality: Dimensions Determining Telepresence. *Communication in the Age of Virtual Reality, 42*(4), 73–93.

Tugendhat, E. (1976). *Vorlesungen zur Einführung in die sprachanalytische Philosophie*. Frankfurt Am Main: Suhrkamp.

United States Department of Defense. (1998). DoD Modeling and Simulation (M&S) Glossary, DOD 5000.59-M. Retrieved March 10, 2007, from http://www.dtic.mil/whs/directives/corres/pdf/500059m.pdf

Wetlesen, J. (1999). The Moral Status of Beings Who Are Not Persons: A Casuistic Argument. *Environmental Values, 8*, 287–323.

Wetzel, L. (2007). Types and Tokens. In E. N. Zalta (Ed.), *The Stanford Encyclopedia of Philosophy (Winter 2007 Edition)*.

Wyller, T. (2000). *Objektivitet og jeg-bevissthet: en aktualisering av Immanuel Kants filosofi*. Oslo: Cappelen.

SECTION II

Philosophical Perspectives on Trust in Online Environments

CHAPTER FOUR

The Role of e-Trust in Distributed Artificial Systems

MARIAROSARIA TADDEO

Introduction

Trust affects many of our daily practises. It is the key to effective communication, interaction and cooperation in any kind of distributed system, including our society and artificial distributed systems (Lagenspetz, 1992). With the diffusion of the Internet and ubiquitous computing, e-trust (trust in digital contexts) has grown and nowadays plays a fundamental role in many of the activities that we perform in virtual contexts, from e-commerce to social networking.

There are at least two kinds of e-trust. The first kind involves human agents (HAs) trusting (possibly a combination of) computational artefacts, digital devices or services, such as a particular website, to achieve a given goal. The users of eBay, for example, trust the website's rating system, and many twenty-first century drivers trust their GPS navigators to indicate the route to follow.

The second kind of e-trust concerns artificial agents (AAs), which develop e-trust in other AAs without the involvement of HAs. Consider, for example, the case of unmanned aerial vehicles such as the Predators RQ-1 / MQ-1 / MQ-9 Reaper. These vehicles are 'long-endurance, medium-altitude unmanned aircraft system[s] for surveillance and reconnaissance missions'.[1] Surveillance imagery from synthetic aperture radar, video cameras and infrared can be distributed in real time to the frontline soldiers and the operational commander, and worldwide, via satellite communication links. The system in the aerial vehicle receives and records video signals and passes them to another system, the Trojan Spirit van, for worldwide distribution or directly to operational users via a commercial global broadcast system. In this case, there are two occurrences of e-trust: that between the artificial system and HAs, and that between the AAs of the distributed system that manage the data flow. This is to say that the Predator's system of data-collecting trusts—at least in senses that will be developed and clarified in what follows, and designated more precisely by the term 'e-trust'—the broadcast systems to acquire the data and transmit them to the users without modifying or damaging them or disclosing them to the wrong users.

In this chapter I focus on the analysis of e-trust relationships that emerge among AAs of distributed systems, like Predators RQ-1 / MQ-1 / MQ-9 Reaper. This analysis has a twofold goal: one, it clarifies the nature of e-trust among AAs and two, it provides a starting point for the elaboration of a model to analyse other, and perhaps more complex, occurrences of e-trust, such as the ones in which HAs are involved.

One reason for focusing on e-trust is that e-trust occurring among the AAs of a distributed system is simpler to describe and model than those in which HAs are involved. This is because trust and e-trust are primarily the consequences of decisions made by the trustor. An analysis of e-trust has to make clear the criteria by which an agent decides to trust, and this is easier to do for AAs than for HAs, because the criteria by which an HA decides to trust are several and heterogeneous (they include economic, attitudinal and psychological factors), and their relevance varies across different circumstances. AAs are computational systems situated in a specific environment, and able to adapt themselves to changes in it. Since AAs are not endowed with mental states, feelings or emotions, psychological and attitudinal criteria are irrelevant. In analysing an AA's decision process, one knows exactly the criteria by which it decides. This is because AAs are artefacts whose criteria are either set by the designer or learned following rules set by the designer.[2]

In the rest of this chapter I will consider AAs that act following a Kantian regulative ideal of a rational agent, and are thereby able to choose the best option for themselves, given a specific scenario and a goal to achieve. I will consider e-trust the result of a rational choice that is expected to be convenient for the agent.

The reader should be alerted that this approach does not reduce the entire phenomenon of trust to a matter of pure rational choice. A system of purely rational AAs, like the one described in this chapter, represents a streamlined and fully-controlled scenario, which turns out to be useful for identifying the fundamental features of e-trust. I agree with the reader who might object that trust involves more than rational choice. I only suggest that we start our analysis of e-trust by considering the simplest occurrences of this phenomenon. We should then use the understanding that such analysis provides to explain more complex scenarios. In addressing the analysis of e-trust we should act like a scientist who plans her experiment. Her first step is to simplify the experimental environment as much as possible in order to focus on the simplest occurrences of the observed phenomenon. She will turn her attention to more complex occurrences only in a second moment, when the simplest ones have been clarified. We should endorse the same minimalist approach; first we shall focus on the simplest occurrences of e-trust and then we shall use the analysis of such occurrences to explain the more complex ones.

I address these issues as follows. In section 2, I examine the foundation of e-trust in trustworthiness. In section 3, I present and defend a new definition of e-trust. Specifically, I describe a method for the objective assessment of the levels of e-trust by looking at the implementation of the new definition in a distributed artificial system. In section 4, I show how the new model can be used to explain more complex occurrences of e-trust, such as those involving HAs. In section 5, I offer some concluding remarks.

Let us now consider e-trust in more detail. I will begin the analysis of e-trust from its foundation: an AA's trustworthiness.

The foundation of e-trust: An objective assessment of trustworthiness

Trust is often understood as a relation that holds when (a) one of the parts (the trustor) chooses to rely on another part (the trustee) to perform a given action, and (b) this choice rests on the assessment of the trustee's trustworthiness. The trustor's assessment of the trustee's trustworthiness is usually considered the foundation of trust and e-trust. Such assessment is usually defined as the set of beliefs that the trustor holds about the potential trustee's abilities, and the probabilities she assigns to those beliefs: see Gambetta (1998) and Castelfranchi & Falcone (1998). Other interpretations have been provided: for example, Tuomela & Hofmann (2003) assess an agent's trustworthiness referring to ethical norms of the context in which the agent performs its actions.

Unfortunately, all these analyses provide a partial explanation of the phenomenon of e-trust and in particular of the definition of trustworthiness, which need to be revised (Taddeo, 2009). In contrast with the claims of these earlier analyses, I have shown that potential trustors do not consider only their subjective beliefs but also objective aspects, such as the results of the performances of the potential trustee. Consider how trustworthiness is assessed in online contexts, for example in online purchases. It has been shown that the potential seller's trustworthiness is assessed on the basis of well-defined and objective criteria (such as previous experiences made with the same website, the brand, the technology of the website, and the seals of approval) that have to be met for users to consider the website trustworthy (Corritore, Kracher, & Wiedenbeck, 2003). This feature of the trustworthiness assessment is well represented in the analysis of the model of a distributed system of rational AAs discussed in this chapter.

In order to determine a potential trustee's trustworthiness, the AAs calculate the ratio of successful actions to total number of actions performed by the potential trustee to achieve a similar goal. Once determined, this value is compared with a threshold value. Only those AAs whose performances have a value above the threshold are considered trustworthy, and so trusted by the other AAs of the system.

The threshold is a parameter that can be either fixed by the designer or determined by the AAs of the system on the basis of the mean performance value of the AAs. Generally speaking, the threshold has a minimum value below which it cannot be moved without imposing a high risk on the trustor. Trustworthiness is then understood as a measure that indicates to the trustor the probability of its gaining by the trustee's performances and, conversely, the risk to the trustor that the trustee will not act as it is expected to do. The minimum value of the threshold is calculated by considering both the risk and the advantage to which a rational AA—where 'rational' is defined within the framework of rational choice theory[3]—might be subject in performing a given action by itself. The threshold must be higher than this value, because for a rational AA it is not convenient to consider as trustworthy an AA that can potentially determine a higher risk and an equal advantage to the one that the AA would achieve acting by itself. Note that the threshold is set higher than the mean value of the AAs' performances. This because the AAs make decisions that, according to their information, will best help them to achieve their goals, and an AA's trustworthiness is a key factor.

Trustworthiness is the guarantee required by the trustor that the trustee will act as it is expected to do without any supervision. In an ideal scenario, rational AAs choose to trust only the most trustworthy AAs for the execution of a given task. If threshold values are high, then only very trustworthy AAs will be potential trustees, and the risk to trustors will be low. This is of course a self-reinforcing process.

There are two more points to note. First, rational AAs assess trustworthiness with respect to a specific set of actions. For this reason, trustworthiness is not a general value: it does not indicate the general reliability of an AA, but its dependability for the achievement of a specific (kind of) goal. It might happen that an AA that is trustworthy for the achievement of a given goal, such as collecting the data about the last *n* online purchases of a user, is not trustworthy for another goal, such as guaranteeing the privacy of the same user.

Second, e-trust does not occur *a priori*. A rational AA would not trust another AA until its trustworthiness had been assessed. Given that trustworthiness is calculated on the basis of the results of performed actions, e-trust could not occur until the potential trustees had performed some action. In systems with only rational agents, no agent whose trustworthiness is not yet measured is trusted. One might consider this as a limit, and object that, since the appraisal of trustworthiness presupposes past interactions, it is impossible to explain the occurrence of e-trust in one-shot interactions, which are indeed very common in the occurrence of e-trust in a distributed system. However, it is important to realize that an agent's trustworthiness is assessed on the agent's past performances and not on its iterated interactions with the same agent. For the assessment of trustworthiness it does not matter whether the trustee performed its actions alone or by interacting with the same trustor, or even another agent. Trustworthiness is like a chronicle of an agent's actions, a report that any agent of the system considers in order to decide whether to trust that agent. Consider the way trustworthiness is assessed in reputation systems, for example in Web of Trust (WOT).[4] WOT is a website reputation rating system. It combines together ratings submitted by individual users about the same trustee to provide community ratings. The resulting collective rating is then shared with all the users who can check it before deciding to trust a particular agent. In this way, any user has the possibility to assess an agent's trustworthiness, even in the case of a one-shot interaction.

On the model I have described, an AA's trustworthiness is a value determined by the outcomes of its performances. The high trustworthiness of an AA guarantees its future good behaviour as a potential trustee and is a justification for the trustor to take the risk of trusting it. Trustworthiness is like the year of foundation written underneath the name of a brand, such as that in Smith & Son 1878. It tells us that since 1878, Smith & Son have being doing good work, satisfying their customers and making a profit, and it is offered as a guarantee that the brand will continue to do business in this way. In the same way, an AA's trustworthiness is a testament to its past successes, which is cited as a guarantee of the future results of its actions. This is the starting point for the analysis of e-trust that I will provide in the next section.

E-trust: A property of relations

Let me begin by presenting a definition of e-trust that is based on the definition of trustworthiness given in the previous section.

> **Definition**: Assume a set of first-order relations functional to the achievement of a goal and that two AAs are involved in the relations, such that one of them (the trustor) has to achieve the given goal and the other (the trustee) is able to perform some actions in order to achieve that goal. If the

> trustor chooses to achieve its goal by the action performed by the trustee, and if the trustor rationally selects the trustee on the basis of its trustworthiness, then the relation has the property of minimising the trustor's effort and commitment in the achievement of that given goal. Such a property is a second-order property that affects the first-order relations taking place between AAs, and is called e-trust.

This definition differs from others provided in the literature, e.g., Gambetta (1998), Castelfranchi & Falcone (1998), and Tuomela & Hofmann (2003), for three reasons. First, it defines e-trust as a second-order property of first-order relations, whereas e-trust is classically (if mistakenly) defined as a first-order relation that occurs between agents. Second, the definition stresses not only the decisional aspects of trust, but also the effects of e-trust on the actions of AAs. Third, the definition highlights the fact that e-trust is goal-oriented. These features deserve some clarification.

Consider, for example, a multi-agents system (MAS) in which AAs interact in commercial transactions.[5] In this case, the AAs of the systems are divided in two categories, the sellers and the buyers. The sellers put on the market their products and the buyers have a set of goods that they need to purchase. A seller (B) and a buyer (A) start the negotiation process anytime that a seller's offer satisfies the buyer's need. The negotiation process might occur with or without trust, depending on the trustworthiness value associated to the seller.[6] According to the definition, when e-trust occurs, there is a first-order relation, purchasing, which ranges over the two agents, and e-trust, which ranges over the first-order relation and affects the way it occurs. In symbols, T (P (A, B, g)), where T is the property of e-trust, which ranges over the relation P, purchasing, occurring between the agents A and B about some good g. Cases like this are usually explained in the literature by arguing that there are two first-order relations, purchasing and e-trust, occurring at the same time. This explanation could be used to argue against the definition of e-trust as a second-order property. Let us accept it for the moment. In the case of the example above, we have a relation P (A, B, g) and a relation T (A, B), where both P and T range over the two AAs. In this case, B is trusted as such, and not merely for the performance of a specific action. From the relation T (A, B) it follows that the buyer trusts the seller to be able to perform any sort of tasks, from honestly selling its goods to guaranteeing the security of the system or to managing the information flaw in the system. This cannot be accepted: it would be like saying that by trusting the eBay seller to sell good products, one also trusts her/him to be a good lawyer or a good astronaut. This cannot be accepted either.

One may argue that A generally trusts B to be an honest seller. If this were so, e-trust would range over all the first-order relations of purchasing that occur between A and B, where B acts, or is supposed to act, honestly. There is a first-order relation between A and B, which occurs every time B acts honestly and which falls in the range of T (A, B). But this simply supports the definition that I provided earlier in this section.

I have said that my definition of e-trust differs from previous ones partly because it takes e-trust to be a property of relations. This means that e-trust affects the way relations occur. The definition states that e-trust affects relations between AAs, minimising the trustor's effort and commitment in achieving a goal. The following example might help to explain this. Consider again the case of the AAs involved in commercial transitions. If the buyer trusts the seller, and the seller is trustworthy, then purchasing occurs more easily than it would have done had the buyer not trusted the seller. The absence of e-trust would force the buyer to spend time browsing the web looking for information about the seller and about its reliability, before purchasing something from that seller.

E-trust relations minimise the trustor's effort and commitment in two ways. First, the trustor can avoid performing an action because it can count on the trustee to do it instead. Second, the trustor does not supervise the trustee's performances without taking a high risk about the trustee's defection. This is a peculiarity of e-trust relations: the absence of supervision is justified there by the trustee's trustworthiness, which guarantees that the trustee is able to perform a given action correctly and autonomously.[7] This minimisation of the trustor's effort and commitment allows the trustor to save time and the energy that it would have spent in performing the action that the trustee executes, or in supervising the trustee. The level of resource-saving is inversely related to the level of e-trust. The higher the e-trust in another AA, the less the trustor has to do in order to achieve its goal and to supervise the trustee. Hence, the more an AA trusts, the less it spends on resources. In this way, e-trust allows the trustor to maximise its gain (by achieving its goal) and minimise its loss of resources.

The inverse relation between the level of e-trust and the level of resource-use falls under the mini-max rule, which is a rule deployed in decision and game theory. The rule is used to maximise the minimum gain or inversely to minimise the maximum loss of the agent making the decision. On the basis of this rule, e-trust and a trustor's resources are variables of the same system, and are related in such a way that the growth of one causes the decrease of the other: hence the notion of a mini-max rule. By using this rule and knowing one of the variables, it is possible to determine the other. The reader should recall that AAs' resources, time and energy, can be objectively quantified. This allows for an objective assessment of the level of e-trust affecting a relation. The model for such an assessment is provided below.

Before considering in more detail the relation between e-trust and the resources of an AA, let us consider the third reason why my definition differs from previous ones. This is that it considers e-trust to be goal-oriented. That e-trust is goal-oriented follows from the teleological orientation of AAs: they act always to achieve some goal, and e-trust is part of a strategy for best achieving those goals. It also follows from the definition in section two of trustworthiness as the ability of an AA to achieve a given goal. We can now look at the model.

An objective model for e-trust levels assessment

The inverse relation between the level of e-trust and the level of resources can be formalised by equating the level of e-trust (y) to the cube of the inverse of the resources (x): $y = (1—2x)^3$. This equation allows one to draw a curve such as the one in Figure 4-1. The curve goes from the maximum level of e-trust, 1, achieved when the level of resources used is equal to 0, to the maximum level of distrust, achieved when the level of the resources used is equal to 1. The power has been introduced to account for the impact of other factors on the occurrence of e-trust, such as the context, the needs of the trustor and the possible urgency of performing an action affected by e-trust. The equation can be modified using different odd powers, depending on how much the extra factors are supposed to affect the emergence of e-trust.

The graph will always show a curve going from the maximum level of e-trust to the maximum level of e-distrust. But, as the reader can see from the graph in Figure 2, the same level of e-trust corresponds to different levels of resources with different powers. More precisely, the power is an expression of the friendliness of the context. The higher the power, the friendlier the context. In a friendly context, AAs

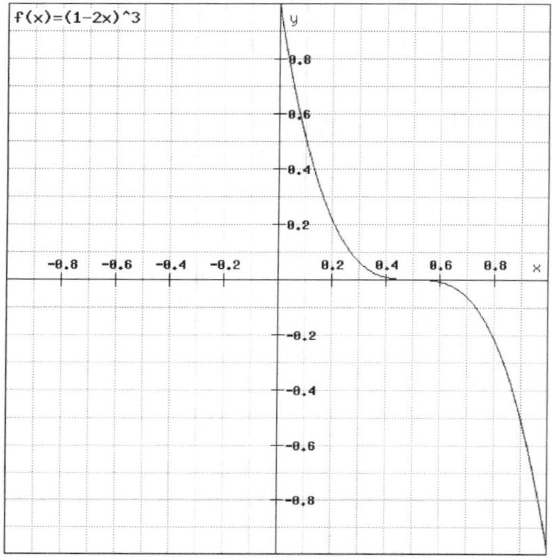

Figure 4-1: Graph showing the levels of e-trust provided by the equation $y = (1-2x)^3$

are less exposed to dangers, and the level of alert and of supervision by an AA of other AAs' performances is lower. The level of resources used to supervise the trustee's performances counts for more in assessing the level of e-trust. For a higher power ($p > 3$), the same amount of resources corresponds to a lower level of e-trust. This model yields both theoretical and practical results.

The theoretical results are as follows. The graph shows that there are three macro sets of e-trust levels: ($y > 0$), ($y = 0$), and ($y < 0$). These sets define a taxonomy of e-trust relations allowing for a distinction between e-trust positive relations ($y > 0$), e-trust neutral relations ($y = 0$), and e-trust negative, or e-distrust, relations ($y < 0$). The presence of e-distrust relation in the taxonomy derived from the model is an important feature, for it shows that the definition defended in this article enables us to explain the entire range of e-trust relations in one consistent framework. E-distrust relations occur when an AA (the dis-trustor) considers the other AA (dis-trustee) untrustworthy and hence uses a high level of its resources to supervise (or ultimately replace) the dis-trustee's performances. Again, as the reader can see from the graph in Figure 4-1, the level of e-trust is lower than 0, and it decreases with the increase of the level of resources used. Hence, e-distrust relations, like e-trust positive and neutral relations, fall under the mini-max rule and are in an inverse relation, as described in section 3.

The practical result is that this method can be implemented by systems for e-trust management in order to determine the level of e-trust occurring in a given relation, and to identify cases of distrust.

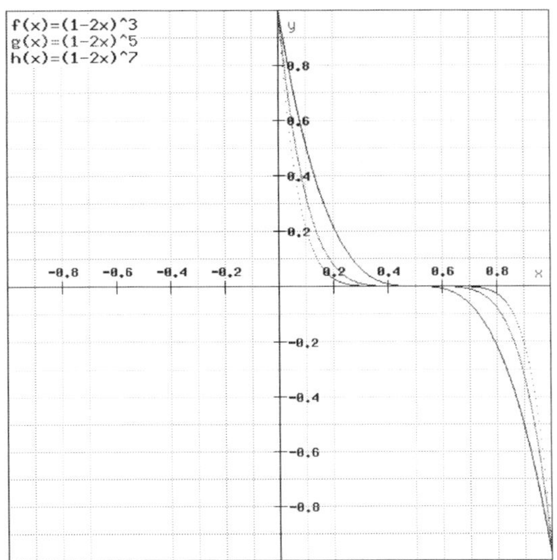

Firgure 4-2: Graph showing the levels of e-trust provided by using the equation $y = (1 - 2x)$ with different powers.

From e-trust to trust

Now that I have described the definition of e-trust in detail, the challenge is to extend it to explain more complex occurrences of trust and e-trust, such as those in which HAs are involved. In this section, I will briefly sketch how the challenge can be met. When HAs are involved, the conditions for the emergence of trust are more complex. There are various criteria, and they change according to the context and to the agents involved. Consider, for example, the case of a rational HA, R, and of a gullible HA, G. Agent R trusts her bank to pay her bills automatically. Her trust in the bank rests on many criteria, such as information in reviews she read about the bank, the good impression she had during a meeting with one of the bank's employees, and the fact that all her bills have been regularly paid so far. R's goal is to have her bills paid and her account properly managed. Now consider agent G, who also trusts his bank to pay his bills, but whose trust rests on a different assessment of its trustworthiness. G chose a bank indiscriminately, and delegated the bank to pay his bills. G's goal is not to have to care about his bills and about the way the bank manages his account. He did not check that the bank would offer a good service. He prefers to take a high risk than be concerned with the trustworthiness of the trustee. Generally speaking, an agent like G prefers to trust, because it allows him to delegate an action and so not to have to care about it.

In the cases of both R and G, there is a first-order relation between an agent and the bank. The relation occurs in the way determined by the presence of trust, i.e., it is qualified by the presence of delegation and the absence of supervision. What distinguishes the two cases is how the trustworthiness of the trustee (the bank) is assessed, and the kind of advantage that is conferred by trusting. Given their heterogeneity, the criteria for the assessment of the trustworthiness of the bank, and the benefits of trusting cannot be specified *a priori.* They must be specified case by case. This is why it is initially rather difficult to extend the definition of e-trust to more complex cases. This difficulty can be overcome if we make the definition and the model more abstract. Trust and e-trust are still defined as second-order properties of first-order relations, able to determine some advantages for the trustor and grounded on the assessment of the trustee's trustworthiness, but the criteria for the assessment of trustworthiness and the benefits of trusting are not specified. Their specification is left to the analysis of the specific cases, while the definition provides a general framework able to explain consistently the occurrences of both trust and e-trust.

By using the methodology of the levels of abstraction (LoA) (Floridi, 2008),[8] it is possible to set the same definition at different levels of abstraction. From a high LoA, the definition does not take into consideration the criteria for the assessment of trustworthiness, but focuses only on the more general aspects of trustworthiness as the foundation of trust and e-trust. The criteria for assessing trustworthiness are specified at a lower LoA. Shifting from higher to lower LoAs is like looking at a map of a city first with the naked eye, and then with a magnifying glass. In the first case, one sees a general model of a city, which would potentially fit any city. This is like choosing a high LoA. If one looks at the map with the magnifying glass, then the map becomes a model of just one city. In the same way, at a low LoA the criteria of trustworthiness and benefits of trusting are specified, and so the definition at this level fits only a specific occurrence of trust or e-trust. At a high LoA the definition is as follows.

> **Definition**: Assume a set of first order-relations functional to the achievement of a goal and that at least two agents (AAs or HAs) are involved in the relations, such that one of them (the trustor) has to achieve the given goal and the other (the trustee) is able to perform some actions in order to achieve that goal. If the trustor chooses to achieve its goal by the action performed by the trustee, and if the trustor considers the trustee a trustworthy agent, then the relation has the property of being advantageous for the trustor. Such a property is a second-order property that affects the first-order relations taking place between AAs, and is called trust.

This definition nowhere specifies the criteria for the assessment of trustworthiness or the benefits of the occurrence of trust in absolute terms. The high LoA guarantees the generality of the definition, which ranges from the cases in which trust is grounded on a pure rational choice to those in which trust occurs on the basis of irrational or random choices. Such a high LoA turns out to be the right level to set a general definition of trust.

Conclusion

There are two aspects that differentiate the analysis presented in this chapter from the analyses of e-trust provided in the literature. The first one is the distinction between the decision-making process performed by the trustor and the occurrence of e-trust itself. On the one hand, the occurrence of e-trust *always* shows the same characteristics, i.e. the delegation of a task and the absence of supervision on the trustee's performances. On the other hand, the criteria that lead the trustor to the choice of trusting another agent *vary* from case to case, depending on the kind of agent performing the decision to trust. For example, a rational AA like the one described in section two decides to trust if and only if the potential trustee satisfies rigid and rationally established parameters. A non-rational agent, like agent G described in section four, decides to trust on the basis of less rigid and less rational parameters. Nevertheless, the presence of trust qualifies the actions of both the rational AA and agent G in the *same way*, that is, allowing the trustor to delegate a given task to the trustee and not to supervise the trustee's performances, determining in this way some advantage for the trustor.

The second aspect is the consideration of both the trustor's decision-making process and the effects of occurrence of e-trust on the trustor's behaviour. The analyses provided in the literature focus on the decision-making process and often disregard the effects of e-trust on the performances of the trustor (see, for example, Castelfranchi & Falcone, 1998).

Within the framework of the analysis presented in this chapter, the decision-making process is only one of the two facets characterising the occurrence of e-trust. The effects of the decision to trust on the trustor's performances are the other facet, which is highlighted by the definition of trust as a second-order property of first-order relations. This definition uses a formal lexicon to describe the role of e-trust as a *shifter*, which alters the way relations occur by facilitating the trustor's task and allowing it to save time and energy.

At the beginning of this chapter, I presented trust as a fundamental feature of social life. In the light of the analysis I have provided, it is clear that trust and e-trust are so fundamental for social relations because the benefits it confers on the trustor invites any agent to get involved in some relationship or in a net of relationships like a social system.

I shall conclude this chapter recalling Hobbes: the advantages arising from mutual trust amongst the agents of a group are the reasons for choosing to end the status of *bellum omnium contra omnes* and establish a society, whether it be real or virtual.

Notes

1. http://www.airforce-technology.com/projects/predator/
2. AAs are computational systems situated in a specific environment and able to adapt themselves to changes in it. They are also able to interact with the environment and with other agents, both human and artificial, and to act autonomously to achieve their goals. AAs are not endowed with mental states, feelings or emotions. For a more in-depth analysis of the features of AAs see Floridi & Sanders (2004).
3. These AAs are assumed to comply with the axioms of rational choice theory. The axioms are: (1) completeness: for any pair of alternatives (x and y), the AA either prefers x to y, prefers y to x, or is indifferent between x and y. (2) Transitivity: if an AA prefers x to y and y to z, then it necessarily prefers x to z. If it is indifferent between x and y, and indifferent between y and

z, then it is necessarily indifferent between x and z. (3) Priority: the AA will choose the most preferred alternative. If the AA is indifferent between two or more alternatives that are preferred to all others, it will choose one of those alternatives, with the specific choice among them remaining indeterminate.

4. http://en.wikipedia.org/wiki/WOT:_Web_of_Trust
5. Such systems are widely distributed in contemporary society: there is a plethora of MAS able to perform tasks such as product brokering, merchant brokering and negotiation. Such systems are also able to address problems like security, trust, reputation, law, payment mechanisms, and advertising. (Guttman, Moukas, & Maes, 1998) (Nwana, et al., 1998)
6. The reader may consider this process similar to the one that occurs in e-commerce contexts where HAs are involved, such as e-Bay for example.
7. Note that 'action' indicates here any performance of an AA, from, for example, controlling an unmanned vehicle to communicating information or data to another AA. For the role of trust in informative processes see Taddeo (2009).
8. In the Theory of Level of Abstraction (LoA), discrete mathematics is used to specify and analyse the behaviour of information systems. The definition of a LoA is this: given a well-defined set X of values, an observable of type X is a variable whose value ranges over X. A LoA consists of a collection of observables of given types. The LoA is determined by the way in which one chooses to describe, analyse and discuss a system and its context. A LoA consists of a collection of observables, each with a well-defined possible set of values or outcomes. Each LoA makes possible an analysis of the system, the result of which is called a model of the system. Evidently, a system may be described at a range of LoAs and so can have a range of models. More intuitively, a LoA is comparable to an 'interface', which consists of a set of features, the observables.

References

Castelfranchi, C., & Falcone, R. (1998). *Principles of Trust for MAS: Cognitive Anatomy, Social Importance, and Quantification.* Paper presented at the Third International Conference on Multi-Agent Systems. (ICMAS'98), Paris, France.

Corritore, C. L., Kracher, B., & Wiedenbeck, S. (2003). On-line Trust: Concepts, Evolving Themes, a Model. *International Journal of Human-Computer Studies 58*(6), 737–758.

Floridi, L. and J. Sanders (2004). On the Morality of Artificial Agents. *Minds and Machines 14*(3): 349–379.

Floridi, L. (2008). The Method of Levels of Abstraction. *Minds and Machines, 18*(3), 303–329.

Gambetta, D. (1998). Can We Trust Trust? In D. Gambetta (Ed.), *Trust: Making and Breaking Cooperative Relations* (pp. 213–238). Oxford: Basil Blackwell.

Guttman, R., Moukas, A., & Maes, P. (1998). Agent-Mediated Electronic Commerce: A Survey. *Knowledge Engineering Review 13*(3), 147–159.

Lagenspetz, O. (1992). Legitimacy and Trust. *Philosophical Investigations, 15*(1), 1–21.

Nwana, H., Rosenschein, J., Sandholm, T., Sierra, C., Maes P., & Guttmann, R. (1998). *Agent-Mediated Electronic Commerce: Issues, Challenges and some Viewpoints.* Paper presented at the Autonomous Agents 98.

Taddeo, M. (2009). Defining Trust and E-trust: Old Theories and New Problems. *International Journal of Technology and Human Interaction (IJTHI)*, 5(2), 23–35.

Tuomela, M., & Hofmann, S. (2003). Simulating Rational Social Normative Trust, Predictive Trust, and Predictive Reliance between Agents. *Ethics and Information Technology* 5(3), 163–176.

CHAPTER FIVE

Trusting Software Agents

JOHN WECKERT

Introduction

Discussions of trust in the virtual or online world have for the most part focused on trust between humans in that medium (for example Nissenbaum, 2001, McGeer, 2004, Pettit, 2004, De Laat, 2005). Can person *A* trust person *B* in that environment in the same way that *A* might trust *B* on the off-line environment? In a previous paper I argued that genuine trust is possible in the virtual world and that this is not surprising (Weckert, 2005). Now I want to focus on different inhabitants of the virtual world, those software agents that are commonly called autonomous. Can we trust these agents or can we at best rely on them? Before answering this question, an account of trust will be outlined (based on Weckert, 2005).

Trust and reliance

The first part of the argument distinguishes trust from mere reliance. If *A* relies on *B* to do *X*, then if *B* does not do *X*, A will be harmed in some way, or at least *A* will be worse off than if *B* had done *X*. I rely on my bicycle to get around, and if the tyres go flat, I am disadvantaged. I rely on the ladder on which I stand to take my weight without collapsing, and if it does collapse I am worse off than if it had not. But in no interesting sense do I trust my bicycle or my ladder. On the other hand, when I rely on my friend to pay back the money that I lent him I do trust him to pay it back. The main difference between these cases is that my friend can choose to pay me back or not; he can choose to help or harm me. In an obvious sense both my bicycle and ladder can either help or harm me, but they do not choose to. So, if *A* trusts *B* to do *X* then *A* relies on *B* making certain choices. While it makes sense in English to say that I trust the ladder, that is a very different sense of trust, and a much thinner one, than the sense in which I trust my friend. Trust involves the trustee having the ability and opportunity to make a choice. But the trustor must also be in a position to choose to trust or not to trust. A beggar may rely on passers-by for sustenance, but have no trust that his needs will be met—he simply has no choice. Trust involves choices; the trustor chooses to rely on the trustee making a particular choice. Reliance, on the other hand, might or might not involve choice.

Choosing to rely on choices in itself is of course not enough. It must be reliance on a choice of a certain kind. Relying on the choice of an enemy does not sound like a good idea. Annette Baier (1986) argues that trust is reliance on a person's goodwill towards one. If *A* trusts *B* then *A* relies on *B*'s goodwill toward *A*; that is, *A* relies on *B*'s choice because of *B*'s goodwill towards *A*. As it stands, this cannot be quite right, as Holton (1994) has shown. First, a confidence trickster might rely on one's goodwill without trusting one, therefore reliance on goodwill cannot be a sufficient condition for trust. It seems that there must be some level of goodwill on the part of the trustor toward the trustee as well. But neither is it a necessary condition, he argues. He rightfully points out that if I trust someone, I do not necessarily rely on that person's goodwill toward *me* (Holton, 1994, p. 65). I could trust someone to care for my children, relying on his goodwill toward *them*, but not necessarily toward me. He might consider me a very poor parent and have no goodwill toward me at all. Therefore, it might be the case that there is always goodwill toward the trustor *or* toward the object of the trust (that is, toward me or toward my children). Holton considers this disjunction but dismisses the view that any goodwill at all is necessary, using one of Baier's own examples. This example is that "we trust our enemies not to fire at us when we lay down our arms and put up a white flag" (Baier, 1986, p. 234). His comment on this is that while we rely on our enemy not to fire, talk of goodwill is out of place, but talk of trust is not. We might trust our enemy here, but we certainly cannot assume any goodwill toward us, he thinks.

There are various possible responses to this. Some goodwill would seem to be possible. The enemy feels good about defeating us, and perhaps some gratitude that we surrendered and did not fight to the death, thus reducing their casualties too. Talk about some degree of goodwill does not seem out of place. On the other hand, it may not always make sense to talk of trust in this situation. Perhaps I do not trust them not to shoot but I do rely on them not shooting because of international agreements and the threat of punishment if they do. More importantly, we may have no real choice but to surrender, in which case too, talk of trust is out of place regardless of whether or not there is any goodwill. This case is not a compelling counterexample against good will but the case is not closed.

Suppose that *B* does not like *A*; he actually dislikes him intensely and extends to him no goodwill at all. But *B* is a very moral person, say in a Kantian, duty-driven sense, and tries hard not to let his feelings get in the way of his doing the right thing. Here it seems that *A* can still trust *B* even in the absence of goodwill. *A* can rely on *B* to make a choice favourable to him (*A*) because *B* is a moral person and believes that this is the moral thing to do. In Aristotelian terms we might say that *A* relies on the fact that *B* is virtuous, and in Kantian terms on *B*'s will being good.

But morality seems not to be the whole story either and in some cases goodwill does appear the key element, just as Baier argues. I might not be able to rely on my friend's morality, but because he is my friend and has goodwill toward me, I can rely on him to make choices that favour me. If there is loyalty amongst thieves and my friend and I are thieves, I may be able to rely on his loyalty to me even though in this case the loyalty does not indicate morality in a broad sense.

In conclusion then, we will say that if *A* trusts *B*, then *A* relies on *B*'s goodwill toward *A* or toward the object of the trust, or *A* relies on *B*'s morality. More generally we will say that if *A* trusts *B*, then *A* relies on *B*'s disposition to behave favourably toward *A*.

Trust as *seeing as*

On the account of trust to be outlined here, it is not the case that a person weighs up the evidence for another's goodwill, morality or disposition on each occasion when trust is an option. Rather, someone is *seen as* trustworthy. If *A* trusts *B*, then *A sees B*'s behaviour *as* being trustworthy, or *A sees B as* trustworthy, or *A* sees *B* as a person to trust. In order not to beg any questions, it can be said that *A* sees *B* as someone who, typically, will keep his word, is reliable, will act with the interests of *A* in mind, will act morally and so on. For the sake of brevity however, we will talk of *A*'s seeing *B* as someone trustworthy. Looking at trust in this way makes it resemble a Kuhnian paradigm (Kuhn, 1970). While not wanting to push this analogy too far, this approach does highlight some of the important features of trust, in particular the intertwining of the views of trust as attitudinal and as cognitive (Baier, 1994), and the robustness of trust. Trust also contains an element of commitment that is often overlooked, although many writers talk about the risk and uncertainty aspects. This is particularly true of cognitive accounts where trust involves weighing up the evidence (for example, Coleman, 1990). While these and other features are not incompatible with accounts couched in terms of attitude or stance (for example, Jones, 1996, Govier, 1997, and Holton, 1994), the *seeing as* approach does make them more obvious. Trust is similar to normal science being conducted within an unquestioned paradigm (Kuhn, 1970). These ideas have something in common too with Lagerspertz, who says that to trust "is to present behaviour in a certain light" (Lagerspertz, 1998, p. 5), in other words, to look at or construe behaviour in a particular way.

One commonly supposed, but rarely argued, feature of trust mentioned already which this model does challenge, is that while trust is difficult to build, it is rather fragile and easy to demolish. (Govier does raise this issue in a slightly different way in the first sentence of her book, where she writes "The human capacity for trust is amazing" (Govier, 1997, p. 3).) The thought is that trust is common even in situations where one would not expect it; where the evidence is against it. That trust is difficult to build is not being questioned here (although it could be), but that it is fragile, is. This view of the fragility of trust seems to be based on something like a Popperian falsificationist view of science. While a theory cannot be verified, an incompatible observation falsifies it. By the same token, it seems to be commonly thought that trust, once violated by the equivalent of a contradictory experience or observation, is broken. We know however, that in science the situation is more complicated than that, and so it is with respect to trust. It is probably true that if I know that *A* has *deceived* me in a situation in which I trusted him, my trust will be weakened or perhaps, extinguished. But most cases are not like this. When expectations are not met, deceit is not always, and probably not usually, involved. Suppose that I trust *A* to do *X* but he does not do *X*. It is unlikely that I will stop trusting him on the basis of one, or even a few, lapses of this kind, unless, of course, I do suspect deceit. There could be many reasons why *A* did not do *X*. There may have been some misunderstanding between us and he did not realise that I expected him to do it. He might have had good reasons for not doing it which are unknown to me, but such that if I did know of them I would approve of his not doing it. I do *not* immediately see *A*'s behaviour as being untrustworthy. In Lagerspertz 's terms, I present *A*'s behaviour in the light of his being trustworthy (Lagerspertz, 1998). Just as in science where, if there is incompatibility between an observation and a theory, there is more than one way to explain the incoherence. One need not reject the theory, or the trust, outright. Quine's *maxim of minimum mutilation* is relevant here (Quine, 1970, p. 7). My most cherished beliefs and attitudes are affected least, or mutilated least, and trust in someone is frequently cherished. I will, if I can, find some explanation that does not involve rejecting my trust.

John Updike illustrates this robustness of trust in his short story "Trust Me" where he writes of a small boy, Harold, who jumps from the side of a pool into the supposedly waiting arms of his father. Unfortunately the father did not catch him, with unpleasant, though not fatal, consequences for the boy. But "Unaccountably, all through his growing up he continued to trust his father." He continued to trust him because he had plausible explanations at hand: "perhaps [he] had leaped a moment before it was expected, or had proved unexpectedly heavy, and had thus slipped through his father's grasp" (Updike, 1987, p. 4). This view of trust is supported by some empirical research as well. According to Rempel, Ross & Holmes (2001), talking of trust in close relationships, "trust can act as a filter through which people interpret their partner's motives" (p. 58). People in high trust relationships interpret their partner's actions "in ways that are consistent with their positive expectations," while those in low trust relationships "are less likely to attribute their partner's behavior to benevolent motives" (Rempel et al. 2001, p. 58).

There will always be some risk and uncertainty associated with trust. If *A* trusts *B*, then *A* takes some risk with respect to *B*, and in a sense, the greater the trust, the greater the risk. I risk a lot more in a loving relationship than I do in a casual friendship. The commitment is greater so there is much more at stake. The *seeing as* account of trust accommodates this, as can be seen in the following discussion of reasonable trust.

When is it reasonable to trust? If "*A* trusts *B*" is cashed in terms of "*A* sees *B* as *Y*," then the reasonableness must be in terms of the reasonableness of the *seeing as*. *A*'s seeing *B* as trustworthy is reasonable to the extent that *A*'s seeing *B* in this way gives a coherent account of *B*'s behaviour. On this view, a few cases of possibly untrustworthy behaviour will not count against the reasonableness of the overall trust. They can be counted as just anomalies to be explained away. *A* will try to "minimally mutilate" his view of *B*. If there are too many anomalies, of course, such a stance becomes no longer viable. This is similar to Kuhn's view of paradigm change in science. Where the threshold is will vary, depending on the strength of the trust. Where the trust is very strong, say the trust in a parent, the trust "paradigm" will be very resistant to challenge. Too much is at stake. Where the trust is less, there will be a correspondingly weaker resistance to challenge.

On this account, where trust is lost, there will be something like a "gestalt switch." After the switch, *A* will see *B* as untrustworthy, and possibly trustworthy actions will be interpreted as anomalies and explained away.

Finally, something needs to be said about the application of this model. So far it has been applied to trust relationships between individuals only, but this requires qualification. Often we trust a person in some respect rather than in all respects. *A* might trust *B* to care for his children, but not to stay sober at the party. Trust is frequently relative to contexts but that is compatible with the *seeing as* model. *A* sees *B* as trustworthy in context *C*, but not necessarily in context *D*. This model of trust can also accommodate trust of institutions. For example, *A* sees the government as trustworthy, or *A* sees business *X* as trustworthy. Again, these could be relativised to particular contexts.

Autonomy

Trust, it has been argued, is a relation between beings that make choices and that deliberate about those choices, that is, autonomous beings. Where there is no ability to makes choices, there can be reliance but no trust. If *A* trusts *B* then both *A* and *B* must be autonomous. This is true even though when *A* sees *B* as trustworthy *A* does not on each occasion of trusting *B*, exercise that autonomy. What is it to be autonomous? An autonomous being, *A*, we will say, is one that is free to make decisions without external constraints, and these decisions emanate from *A*. If there is a choice between two courses of action, *X*, and *Y*, then *A* is autonomous to the extent that he can freely choose either of them. The position taken here is a compatibilist one and a form of agent causation, although probably not the most common form (but see Markosian, 1999) *A* is free to choose but the decisions that he makes are caused; they are caused by him. There is a sense of course in which the causal chain does not start with him alone. What he is, is a result of previous events; his parents' genetic make-up, his early education and training, and his environment. So his decisions are caused by combinations of these factors, but they are real decisions for all that. The alternative view is that some things are not caused, but that counts against autonomy more than causation does. If *A* chooses *X* rather than *Y* and the reasons for his decision were not caused by anything, then he is hardly autonomous but rather a helpless pawn of randomness. *A* is what he is because of a multitude of factors, including his genetics, his hard-wiring, and what he is causes him to make particular decisions. Some might conclude from this that we delude ourselves when we think that we can make genuine choices or decisions but that is not an easy position to defend. We certainly feel as if we deliberate and often we agonise over our decisions. There are various reasons for our deliberations over decisions. Some are obviously because of ignorance. *A* wants to buy shares but does not know which ones are more likely to rise. His decision will depend on the advice that he gets and on his attitudes to the sources of the advice (specifically, to what extent he trusts the advisors), and partly on the extent to which he is a risk-taker. In other cases we have a clash of desires. I want to go walking in the mountains but I also want to watch the football match and unfortunately cannot do both. A third type of decision must be made when there is a clash between a desire and a duty. I want to go walking but know that I ought to finish writing the paper by the end of the week, as I had promised to do. In all of these cases we must genuinely decide what course of action to take, even though our decision will ultimately be a result of what we are, our character, and what we know or believe at the time of the decision. If we had perfect rationality and complete knowledge, and no desires we might not have to deliberate over our decisions, but we are not like that.

This short outline of autonomy assumes that there is a causal story to tell for each decision that someone makes, although normally we do not bother to try tell it and in most cases could not anyway, simply because such a story would be far too complex. If I choose to go walking in the mountains rather than finishing a paper that is already late, this can be explained by various facts: I like walking more than writing, I'm basically lazy, I'm not very conscientious, and so on. And each of these facts can be explained in terms of my genetic make up, my upbringing, my environment and experiences, and so on. I made the decision because of the way that I am and the way that I am is a result of the factors mentioned. So there is a sense that my decision was caused by those factors. But this in no way rules out my being autonomous and fits in closely with our common ways of talking. We talk of people doing certain actions because they are kind people, and praise them for this. The fact that someone had kind parents and was raised in a loving and secure environment does not make us think less of him or her *considered as* a kind person. Likewise, we blame a person for having a violent temper even if we know that he or she comes from a long line of bad-tempered people.

There is obviously much more to say about autonomy and about compatibilism, for example, the role of second-order desires (Frankfurt, 2005) and the situation of people who act under the influence of drugs or with some brain disorder (see for example Levy & Bayne, 2004). These things complicate the fairly simple account above but they do not add anything substantial and do not show that that account is wrong. Hence I will not further defend this claim here.

Software agents

The next part of the argument must consider whether or not so-called autonomous software agents are the sorts of things in which trust can be placed; does it make sense to say that they can be *seen as* trustworthy? First, what is an agent in general and second, what is the relevant kind of software agent?

In philosophy an agent is someone who does something or who can or will do something or has done so in the past. A moral agent, for example, can be morally praiseworthy or blameworthy for his or her actions. Another sense of agent is someone who does something for someone or acts on that person's behalf. My travel agent organises my travel, for example. While in this sense of agent, an agent's overarching goal, *qua* agent, must be to satisfy the goal of the person for whom he is an agent, the agent has autonomy within certain parameters. In order to save me time and effort, he must be free to make decisions and judgements. So the agent is the sort of being that it makes sense to trust. I will normally choose as an agent someone who I believe that I can trust. This latter kind of agent is also an agent in the first sense.

A software agent is also something that can act on someone's behalf in order to free his or her time for other activities or to undertake tasks for which the person lacks the relevant knowledge or skills. This is the reason for the development of such agents. For this, some level of "intelligence" is required. These agents must also have a degree of autonomy:

> An agent is a software component capable of flexible, autonomous action in a dynamic, unpredictable and open environment. Teams of agents can interact to support distributed models of problem-solving and control. Multi-agent systems can be effective in accommodating and reacting usefully to environmental changes. Agent technologies are a natural extension of current component-based approaches, and have the potential to greatly impact the efficiency and robustness of the software engineering process. Research into agents is one of the most dynamic and exciting directions in computing today. (Agentlab, 2009)

One example of such an agent is *Commonsense investing: Bridging the gap between expert and novice*, an "intelligent personal-finance advisory agent" being developed at MIT by Ashwani Kumar and Henry Lieberman, which

> bridges the gap between the novice user and the expert model of the finance domain. The agent uses common-sense reasoning and inference for associating the user's personal life, financial situation, and goals with the attributes of the expert domain model and vice versa. (MIT, 2005b)

Another, by Hugo Liu and Henry Lieberman, is ARIA (Annotation and Retrieval Integration Agent),

> ...a software agent that acts as an assistant to a user writing email or Web pages. As the user types a story, it does continuous retrieval and ranking on a photo database. It can use descriptions in the story text to semi-automatically annotate pictures based on how they are used. (MIT, 2005b)

It must be noted that it is not claimed that these agents are fully autonomous, rather they are called "semi-autonomous" (MIT, 2005a). Current agents probably all could be so described given that their autonomy is within fairly narrowly circumscribed boundaries, but that does not affect the argument here and we will consider autonomy more closely shortly.

Are software agents the kinds of things that can be trusted? The software agents of concern here are those that are commonly now called autonomous agents:

> An *autonomous agent* is a system situated within and a part of an environment that senses that environment and acts on it, over time, in pursuit of its own agenda and so as to effect what it senses in the future (Franklin & Graesser, 1997, 26).
>
> A good internet agent needs these same capabilities. It must be *communicative*: able to understand your goals, preferences and constraints. It must be *capable*: able to take options rather than simply provide advice. It must be *autonomous*; able to act without the user being in control the whole time. And it should be *adaptive*; able to learn from experience about both its tasks and about its users preferences. (Hendler, 1999)

Relating definitions such as these to more familiar human agents, Ndumu and Nwana write:

> Essentially therefore, a software agent is any software entity to which can be ascribed similar attributes implicit in the everyday usage of phrases such as "travel agent" or "estate agent." That is, among other things, goal-orientated behaviour, knowledge of the problem-solving techniques of their domain, autonomous and pro-active functioning on behalf of a customer, and the ability to learn of dealing with a customer or a particular problem. (Ndumu and Nwana , 1997, 3)

The core element for the purposes here in these definitions, and others in the literature (e.g., Wooldridge, 1997), is autonomy. Autonomy is central because it plays a vital role in the distinction drawn earlier between trust and reliance. If talk of trust with respect to software agents is to make sense, those agents must have autonomy.

It has been argued that *A* trusts *B* to do *X* if *A* chooses to rely on *B* choosing to do *X*, perhaps because *B* is moral or because *B* has goodwill towards *A*. The question now is, does it make sense to talk of placing trust in online agents? To be something that can be trusted an agent must be able to make choices in some significant sense; it must be autonomous. We will say that a software agent is autonomous if the following conditions hold:

1. the agent has a memory that stores information as a result of both contact with its environment (inputs) and past inferences and whose content changes over time as a result of further environmental inputs and continuing inferences;
2. the agent can make inferences based on the contents of its memory and the information that it receives from its environmental inputs;

3. the agent has some initial internal structure, parts of which can change as the agent learns from its environment and from past inferences;
4. the agent makes decisions based on its inputs and its internal state, including its memory.

Clearly an agent must also have desires to do certain things rather than others and goals that it tries to achieve. These can be stored in its memory or hard-wired in its internal structure.

That a software agent can satisfy 1–4 is easy to show. First, memory is an essential component of just about any computer system, and memories do change, both as a result of new data coming in through input devices and through inferencing and calculation. Second, all programs make inferences, most commonly through straightforward deduction, usually *modus ponens*, but other forms, for example induction and abduction, are also possible. Third, computer systems have both a physical structure that is hard-wired, and some initial software program. While the hard wiring cannot be changed during the execution of the program, the software can be designed so that it can. In computer languages such as Lisp and Prolog, there is no essential difference between the data and the executable code, so just as the data stored in memory can be changed, so can the code itself. Software that can change its code as it executes clearly learns in some sense not too different from the sense in which we learn. Over time the same inputs will produce different outputs. Finally, computer systems constantly make decisions, usually by the use of *If... Then... Else* structures or some variant like *Case* statements. Software agents are a type of computer system so they too have these capacities.

Agents, autonomy and trust

Are these conditions as they are instantiated in a software agent sufficient for autonomy? An autonomous human must have a memory, be able to interact with his environment, to reason, and to learn both from his environment and as a result of reasoning. Such a human can make decisions that emanate from him alone, and if there are no external constraints on him, he is autonomous. Similarly, a software agent can make choices and the sense in which it makes choices is not too different from the sense in which we make choices. Certainly at any point in time its choices depend on its internal state and its inputs, but the same is true of us. My choices are not random; they are a result of my internal state, my character, and the inputs that I am receiving from my environment. The software agent's choices are a result of its internal state, its structure and its inputs from its environment.

Two immediate objections are that (1) we deliberate in a way that a software agent cannot, and (2) that we can explain how we arrived at our decisions. The second objection can easily be answered. Those computer systems known as expert systems have long had the capacity to explain how they arrived at a particular decision, so there is no reason why an agent would not be able to do the same. The first objection too can be answered. What do I do when I deliberate (at least ideally)? I look at the evidence and assess it, compare various pieces of evidence, examine my goals, desires and values, and the soundness of my reasoning. There is no reason in principle why a software agent could not also do all of these. The literature on agents and on artificial intelligence abounds with examples of such systems, albeit simple ones compared with humans, but there is no reason to doubt that agents as sophisticated at deliberating as humans are possible, certainly in principle but probably also in practice.

Nevertheless it does seem as if the deliberations that I make are caused by me in a different sense from the way in which the decisions that a software agent makes are caused that agent. Certainly my genetic makeup, my education and environment and my goals play a part in my decisions, but I, whatever *I* am, can override all of these, and it is in this that my autonomy consists and what makes it different from the autonomy of software agents. While this sounds like a relevant difference, it is not clear that it helps. If it means that I can override certain desires in favour of others or in favour of some duty, that is nothing more than we have already seen and reveals no interesting difference. Some desires have precedence over others and on occasion, over some duties. If it means that I override all desires, goals and so on, it is not clear that I am acting rationally. If what *I* decide is not based on my genetics, environment or learning, things that determine my character, it is unclear on the basis of what *I* make the decision. That is hardly a sign of an autonomous being, at least not of a desirable one, and certainly not one deserving of trust. It is more a sign of someone who decides randomly. We could not rely on such a person choosing to act in our interests either because of his morality, goodwill or disposition. This line of argument, then, for showing a relevant difference between humans and software agents with respect to autonomy is not promising.

A further objection is that only a conscious being can deliberate and therefore software agents, not being conscious, cannot deliberate and so are not autonomous. This has intuitive appeal. Conscious beings deliberate over their decisions whereas non-conscious ones can make decisions only mechanically. A full response to this would involve an in-depth discussion of philosophy of mind and consciousness but here a few comments designed to force us to question our intuitions will have to suffice. First, what is our evidence that consciousness is necessary for deliberation? I know that when I deliberate that I am conscious, but I have no such firsthand evidence for anyone else. Given that other people are more or less like me, I reasonably assume that they consciously deliberate too. In most cases there is no reason to doubt it. Software agents are not much like me so it is much more difficult to seriously entertain the view that they can consciously deliberate. Furthermore, they are made by someone and the principles for their construction are understood. They can make decisions but only mechanically, our intuitions tell us. But some care is required here. A so-called "autonomous agent" makes decisions on the basis of not only its initial state but also on the basis of its past experience from which it has learned. This is not so different from us. Still, our being different in relevant ways is a strong intuition but not too much store should be placed on that. Intuitions are not knock-down arguments. A conflict between an intuition and a good theory is not enough in itself to refute the theory. Intuitions are not necessarily reliable. They vary between different people and change over time. I may well in the future have the intuition that certain computer systems have consciousness. We rely on them to make choices favourable to us. We see them as trustworthy. Perhaps we will see them as conscious as well.

If the preceding argument is correct, at least a *prima facie* case can be made that we can place trust in agents and not merely rely on them. We can place trust in them because they are autonomous in a significant sense.

Earlier, trust was spelt out in terms of reliance that emphasised moral choices and goodwill. This implies that if we are to trust software agents, they must be moral entities capable of making moral decisions and having goodwill. While I do not want to rule out this possibility, neither do I want to be committed to it here. Therefore instead of talking of trust in terms of reliance on moral choices or good will, we will change the emphasis and say that *A* trusts *B* if *A* relies on *B*'s *disposition* to make choices favourable to *A*. This leaves it open whether or not *B* is a being capable of morality or goodwill. If *B* has a disposition to make choices that are in *A*'s interest, that is enough. This way of talking about autonomous agents is not forced. We do talk of people having happy or helpful dispositions, and these dispositions play a causal role in the actions that they choose.

Conclusion

In this paper it has been argued that trust is a certain kind of reliance, reliance on certain choices being made. Autonomy is central to this account, both autonomy of the trustee and the trustor. It was also argued that software agents can be autonomous in much the same way as humans can be. A consequence of this is that autonomous software agents can be trusted in essentially the same sense in which humans can be trusted. A software agent can be seen as trustworthy. This is nowhere near a complete account of trust in general nor of what is involved in trusting online, autonomous software agents, but if correct, it allows for models of trust richer than trust as mere reliance, to be applied to these agents.

References

Agentlab. (2009). Retrieved from http://www.agentlab.unimelb.edu.au/publicity/agentlab-brochure.pdf (accessed 11 May 2010).

Baier, A. (1994). Trust and its vulnerabilities. In A. Baier (Ed.), *Moral Prejudices: Essays on Ethics* (pp. 130–151). Cambridge: Harvard University Press.

Baier, A. C. (1986). Trust and antitrust. *Ethics, 96*, 231–260.

Coleman, J. S. (1990). *Foundations of social theory.* Cambridge: Harvard University Press.

De Laat, Paul B. (2005). Trusting virtual trust, *Ethics and Information Technology, 7*, 167–180.

Franklin, S., and Graesser, A. (1997). Is it an agent, or just a program?: A taxonomy for autonomous agents. *Intelligent Agents III Agent Theories, Architectures, and Languages,* Lecture Notes in Computer Science, Volume 1193 (pp. 21–25). Berlin / Heidelberg: Springer.

Frankfurt, H.G. (2005). Freedom of the will and the concept of a person. In *The Importance of What We Care About* (pp. 11–25). New York: Cambridge University Press. Reprinted from *Journal of Philosophy*, LXVIII, January 14, 1971)

Govier, T. (1997). *Social Trust and Human Communities*. Montreal: McGill-Queen's University Press.

Hendler, J. (1999). Is there an intelligent agent in your future? *Nature Web Matters.* Retrieved from http://www.nature.com/nature/webmatters/agents/agents.html (accessed 11 May 2010)

Holton, R. (1994). Deciding to trust, coming to believe. *Australasian Journal of Philosophy, 72*, 63–76.

Jones, K. (1996). Trust as an affective attitude. *Ethics, 107*, 4—25.

Kuhn, T. (1970). *The Structure of Scientific Revolutions.* Chicago: University of Chicago Press.

Lagerspertz, O. (1998). *The Tacit Demand.* Dordrecht: Kluwer.

Levy, N. and Bayne, T. (2004). A will of one's own: Consciousness, control, and character. *International Journal of Law and Psychiatry, 2*(7), 459–470.

Markosian, N. (1999). A compatibilist version of the theory of agent causation. *Pacific Philosophical Quarterly, 80*, 257–277

McGeer, V. (2004). Developing trust on the Internet. *Analysis and Kritik, 26*, 91–107.

MIT. (2005a). MIT Media Lab, Software Agents Group. Retrieved from http://agents.media.mit.edu/ (accessed 11 May 2010)

MIT. (2005b). MIT Media Lab, Software Agents Group, Projects. Retrieved from http://agents.media.mit.edu/projects.html (accessed 11 May 2010)

Ndumu, D.T., and Nwana, H.S. (1997). Research and development challenges for agent-based systems. *IEE Proceedings Software Engineering, 144*, 1, 2–10.

Nissenbaum, H. (2001). Securing trust online: Wisdom or oxymoron? *Boston University Law Review, 81*, 101–131.

Pettit, P. (2004). Trust, reliance and the Internet. *Analysis and Kritik*, 26, 108–121.

Quine, W. V. (1970). *Philosophy of Logic.* Englewood Cliffs, NJ: Prentice-Hall.

Rempel, J.K., Ross, M., and Holmes, J.G. (2001). Trust and communicated attributions in close relationships. *Journal of Personality and Social Psychology* 81, 57–64.

Updike, J. (1987). *Trust Me: Short Stories.* New York: Knopf.Weckert, J. (2005). On-line trust. In R. Cavalier (ed.), *The Impact of the Internet on Our Moral Lives* (pp. 95–117). Albany, NY: SUNY Press.

Wooldridge, M. J. (1997). Agent-based software engineering. *IEE Proceedings Software Engineering,* 144, 26–37.

CHAPTER SIX

Trust in the Virtual/Physical Interworld

Annamaria Carusi

Background

The borders between the physical and the virtual are ever-more porous in the daily lives of those of us who live in Internet-enabled societies. An increasing number of our daily interactions and transactions take place on the Internet. Social, economic, educational, medical, scientific and other activities are all permeated by the digital in one or other kind of virtual environment. Hand in hand with the ever-increasing reach of the Internet, the digital and the virtual go concerns about trust. In the increasing numbers of cross-disciplinary attempts to understand the way that the Internet is changing our societies, 'trust' is a truly cross-boundary word, used just as frequently by computer scientists as it is by economists, sociologists and philosophers. Concerns in the name of trust are articulated about the objects and artefacts found, accessed or bought on the Internet, about the people with whom we interact on the Internet, and about the technological systems and infrastructures that enable us to carry out activities of different types.

Much of our preoccupation with trust is brought about by the technologies for so-called virtuality themselves. New technologies are disruptive of established practices, including those involving existing technologies, and so force us to raise issues about trust that we would otherwise not raise (e.g., of the telephone). This is a non-trivial point. It is not just that new technologies force us to contemplate trust in such a way that there are the technologies on one side and the question of trust on the other. When I contemplate whether I can trust the branch of a tree that has fallen over a small stream to bear my weight if I use it to cross the stream, the branch of the tree does not disrupt my trust practices, but relies on them; more importantly, it does not change my practices—of walking, of trusting the ground (and other things) I step onto. However, the technologies of virtuality that we are now immersed in *do* change our practices—of doing things like buying books, accessing data and interacting over them. They change our practices in such a way as to push trust into the fore (where it is often simply in the background). When trust occupies this foregrounded position, it also—at the same time and by the same token—brings about a more reflective stance, where reasons for trusting or not are actively and consciously sought, where in buying a book from a store they would not be. Often, what appears to be a more rational stance is taken in virtual environments. Thus for example, buying a book on Amazon and its associated sellers involves going through a number of 'trust' steps, including considering the ratings of the booksellers. These ratings bring to the fore a reason to trust (based on the experience of other buyers) and apparently give it a 'rational' form: a quantitative measure of the positive/negative experiences of other buyers. The technical ease with which ratings are instituted means that a particular construal of a reason for trust, in terms of a quantitative measure, is pushed to the fore, and thereby your trust is 'rationalised'—in the sense of making something appear rational which is not necessarily so. This is not a step that one would take when buying a book from a store, even though there may have been at the back of one's mind a recommendation from a friend, or a review read in the local newspaper. On Guido Mollering's interpretation of Simmel, trust relies on an inductive leap based on weak evidence and much faith (Mollering, 2001): applying this to the rating system used by many online vendors, we might say that the rating system tries to make it appear that our inductive leaps are based on stronger evidence and less faith, but in so doing, it has also re-shaped the range of what is taken to count as a reason for trust. In this way, the overall landscape of the way in which we trust is changed; what was once a (mole-) hill, has become a mountain.

Two domains in which the demand for reasons re-shapes rather than reinforces existing trust practices are those of security online and those of discerning which online objects to choose out of a plethora of choices.[1] In both of these domains it would seem that we are entitled to demand reasons for trusting. Thus when I try to enter a secure site, it is demanded that I should identify myself in some way or another, and increasingly complex systems of encryption and identification are designed and developed, sometimes at the risk of making the system virtually unusable. The assumption is that the authentication of the identity of users is a reason for the system to trust users before allowing them access. The language used underwrites this inversion: the system must trust users, so that users can trust the system (and the other users on it). Similarly, there is a demand that objects we find online—music, texts, information—be authenticated in some way, give us reasons to trust them in advance of our interactions with them. It is irresponsible to underestimate the very real challenges that designers and developers of systems face, and none of us would like to risk someone hacking into our online bank account and emptying it out. However, it is also true that a particular rhetoric and mode of framing the challenges, risks and responsibilities are associated with online trust.

This paper reflects on the implications for trust of the way we shape our technologies and they in turn shape us, for example, in the way we trust and the extent to which we can trust ourselves as trusters. The account I am working towards is an ecological and co-evolutionary view of trust and technologies, which attempts to hold in view the complex inter-relationships between the agents and other entities within and across environments.[2] First, I consider the ways in which problems of justifying trust are analogous to problems of justifying knowledge, and claim that trust, like knowledge, cannot be justified from an external position. Second, I outline an account of internal relations drawn from phenomenology. This is followed by a discussion of three aspects of trust which are internally related to it: value, reason and morality. Following from these considerations, I claim that the distinction between virtual and physical, or virtual and real objects cannot be upheld. Along the way, I draw upon a number of different examples of activities undertaken online, including those on commercial sites, or so called social networking sites, but in particular those which relate to undertaking scientific research of different kinds. These are interesting examples of environments where much is at stake for the parties involved, and where trust plays itself out alongside the science. I also draw upon a rather eclectic range of philosophers, from different traditions of philosophy and different philosophical interests. My concern is not with being a purist, but with drawing on a variety of resources which point towards an account of trust where it is recognised as not an isolated final act, predicated upon but independent from underlying support, but as deeply embedded in the social-technological configuration in which it plays its essential role.

Trust and knowledge

Firstly, I would like briefly to draw a parallel between trust and knowledge, and the demands that it is possible to make on each of these. For many hundreds of years over the history of epistemology, a favourite occupation was that of trying to beat the sceptic about knowledge, for example, with respect to knowledge about the external world, knowledge of other minds, and so on. Now, we know that it is very hard to beat the sceptic, and that (arguably) perhaps the sceptic cannot be beaten, though there have been some spectacularly good moves made in the game. The best rebuttal of the sceptic is to refuse the game; in fact, more than this, to point out that the game makes no sense. To point out, for example, along with Hume and Wittgenstein, in their own very different ways, that there is no place from which to stand in order to shout out the sceptic's challenge. As Peter Strawson (1985, 19–20) has pointed out:

> ...we can, I think, at least as far as the general sceptical questions are concerned, discern a profound community between [Wittgenstein] and Hume. They have in common the view that our 'beliefs' in the existence of body and, to speak roughly, in the general reliability of induction are not grounded beliefs and at the same time are not open to serious doubt. They are, one might say, outside our critical and rational competence in the sense that they define, or help to define, the area in which that competence is exercised. To attempt to confront the professional sceptical doubt with arguments in support of these beliefs, with rational justifications, is simply to show a total misunderstanding of the role they actually play in our belief-systems. The correct way with the professional sceptical doubt is not to attempt to rebut it with argument, but to point out that it is idle, unreal, a pretence; and then the rebutting arguments will appear as equally idle; the reasons produced in those arguments to justify induction or belief in the existence of body are not, and do not become, *our* reasons for these beliefs; there is no such thing as *the reasons for which we hold* these beliefs. We simply cannot help accepting them as defining the areas within which the questions come up of what beliefs we should rationally hold on such-and-such a matter.

I would like to take my lead from Strawson, and to suggest that there are some demands on trust that it makes no sense to make, that there is no position from which to make them. Just as the sceptic about the external world or about other minds will be satisfied only with a justification of beliefs that stands outside of those beliefs and the ways they are formed, some accounts of trust will be satisfied only with a justification of trust in the light of reasons (evidence, signs, a rational calculation) that stand outside of the trust relationship in some way, is other to it. Once again, Hume is a good example, but this time pulling in the opposite direction (though here Hume is talking about trust, not knowledge), as his insistence is that trust in the testimony of others (as opposed to our broad epistemological framework) should ultimately be based on observation; that is, an epistemic item which is itself external to trust, one from which all trust is eliminated by simply not being required:

> Thus we believe that CAESAR was kill'd in the senate house on the *ides* of *March*; and that because this fact is establish'd on the unanimous testimony of historians,...and these ideas were either in the minds of such as were immediately present at that action...or they were derived from the testimony of others, and that again from another testimony...'till we arrive at those who were eye-witnesses and spectators of the event. 'Tis obvious all this chain of argument or connexion of causes and effects, is at first founded on those characters or letters, which are seen or remember'd and without the authority either of the memory or senses our whole reasoning wou'd be chimerical and without foundation. (Hume 1902, 83)

I suggest that we should align trust with Hume's stance on the limits of knowledge rather than with his stance on testimony, and further, that this demand for externality cannot be demanded of trust, because trust functions and can only function in a way that we can trust, within a system of internal relations between and among trusters and what they trust.

Internal relations

In an external relation, the items related are what they are independently of one another.[3] To use an example from Sartre: a lamp and a book are what they are independently of each other. It takes an observer to bring them into relation with one another: 'The lamp is to the left of the book'. If two things are internally related, neither would be what it is were it not for the other. Thus for example, I would not be what I am were it not for my relations to other beings capable of saying 'I'; my future is not what it is were it not for its relation to my past; the colour of a perceived red carpet is not what it is except in relation to its texture (a 'fleecy red' or a 'woolly red'); a perceived object is not what it is (as perceived object) except in relation to actions, activities, sensations and emotions. For the last-named examples, the examples given by Merleau-Ponty (1962) of the ways in which internal relations between perceptions and actions form the phenomenology of perception are particularly striking (and will be of special relevance for questions of trust):

> A wooden wheel placed on the ground is not, *for sight*, the same thing as a wheel bearing a load. A body at rest because no force is being exerted upon it is again for sight not the same thing as a body in which opposing forces are in equilibrium. The light of a candle changes its appearance for a child when, after a burn, it stops attracting the child's hand, and becomes literally repulsive. (1962, 52)

I explore more fully in the next section how value, reason, and morals (understood in specific ways), as three aspects of trust, are internally related to the trust they are meant to support. This will make clear that the demands made on trust cannot be made except by paying attention to these internal relations. Without recognising the internal relationships of the relata in the trust relationship, they tend to come apart. Once apart, they cannot be put together again in a way that can satisfy the demand for a justification of trust, for example, in a way that the evidence for an act of trust will ever be sufficient to justify the trust. To see this, first we need to turn to a consideration of some of the central characteristics of the act of trusting.

Three aspects of trust

Which parts/aspects are related, and how, is seen differently in different accounts of trust. The three aspects of trust I consider are value (interests), reason (evidence, effectiveness), and morals (trusting the goodwill of the trustee in some way or another). How are these three aspects of trust internally related?

Value

Annette Baier (1986) suggests that the structure of trust is reflected in a three-place relation: X trusts Y with some valued thing, where the value of the item in question is to be understood in the broad sense of instrumental, aesthetic, epistemic, or moral value. This suggests that the valued thing is external to the act of trust. However, this is an oversimplification for many reasons, but one important reason is that there is never, in a situation where trust is at play, just one valued thing. There are many intermediate values, and values related to each other. In addition, intermediate values can become valued in themselves, or shift in the scale of mediacy. For example, science and research are increasingly digitised and available through Internet-enabled media. On the one hand, this creates a crisis of trust in that the much greater quantities of data and other items available, much of it without the mediation of the print publication structure and institution, create doubts about which items can be trusted. Thus, the value of any individual item is put into question. On the other hand, with digital data repositories of all kinds becoming increasingly common, the digital data set is beginning to be a value in its own right for scientists and researchers, where once data were merely a means to a greater value: the publication.[4] This shifting of values up and down the scale is not independent of the technologies for exchanging the items that we (scientists, readers, and users of the systems generally) care about; their value to us is what it is in part because of the ways in which they are positioned by the technologies. Similarly our own values (in the broad sense) are not independent of the technologies: to value a data set is also to value a different kind of epistemic attribute than what is valued in a publication.

Reason

Several writers in the area of trust insist on the role of evidence for trust to be trustworthy (for example, in very different ways, Gambetta (1994), Goldman (1999) and Pettit (1995). In the Human Computer Interaction literature on trust, great store is set on what counts as a sign of trustworthiness for users (for commentary on this see Riegelsberger, Sasse, & McCarthy 2005). However, trust, apart from evidence, also always implies a leap (not unlike the inductive leap of faith that so troubles many epistemologists). How big you think this leap is will depend on how good you think human beings are at seeking and evaluating evidence—not something I wish to broach directly here.[5] However, even if they are moderately or even very good at this, there is always a leap to be made. In this context, the work of Guido Mollering is of interest. Mollering draws upon the ideas of sociologist and philosopher Georg Simmel in not regarding trust merely as a process of making inferences about trustworthiness from prior observations. Trust, he writes, is not simply a matter of projecting onto the future, evidence gained from past experience. Simmel claims that there is a further element which he describes as 'socio-psychological quasi-religious faith'. Mollering writes:

> In another source, interestingly right after a short discussion of lying, Simmel (1950 [1908], 318) describes trust as 'an antecedent or subsequent form of knowledge' that is 'intermediate between knowledge and ignorance about a man.' Complete knowledge or ignorance would eliminate the need for or possibility of trust. Accordingly, trust combines weak inductive knowledge with some mysterious, unaccountable faith: 'On the other hand, even in the social forms of confidence, no matter how exactly and intellectually grounded they may appear to be, there may yet be some additional affective, even mystical, "faith" of man in man'. (Mollering, 2001, 13)

This faith, in turn, is expressed as 'the feeling that there exists between our idea of a being and the being itself a definite connection and unity, a certain consistency in our conception of it, an assurance and lack of resistance in the surrender of the Ego to this conception, which may rest upon particular reasons, but is not explained by them'. (Simmel (1990 [1907], 179) quoted in Mollering, 2001, 405–6)

On this account, trust involves both ignorance and knowledge. However, the ignorance is a kind of wilful ignorance. Mollering sees this as a form of suspension:

in trusting we suspend (or I would say bracket) our vulnerability and uncertainty. It is there and not there; it could be useful to liken it to our suspension or bracketing of disbelief in fiction. On Mollering's account, the framing for this—what makes it possible—is the moral aspect of trust, the obligation made on the other to behave in accordance with the trust placed in him or her, and the consequences for not doing so (See Mollering, 2001 and 2008).

This moral aspect of trust is possibly best introduced via a parallel drawn between trust and bad faith. The situation described by Mollering, following Simmel, is structurally similar to that of self-deception. Sartre gives a famous account of self-deception as a form of bad faith in the examples of the waiter who insists on becoming wholly a waiter and the woman who refuses to acknowledge that she is allowing herself to be seduced (Sartre, 1958, 59ff.) There are myriad examples in our ordinary daily lives. According to the picture of bad faith given by Sartre, bad faith is paradoxical because it involves one and the same person both knowing and not-knowing something. Sartre's solution to this paradox is to see the person as not self-identical; that is, the self is both its facticity (that is tethered to its history, environment and situation) and a transcendence thereof (that is, free; anything is possible for the self). Radical freedom causes anxiety, even anguish, and is denied in an attempt to escape it.

The point I want to insist on is how this is actually possible. According to the account put forward by Katherine Morris (2008, 86):

> [...] a lie is intentional, and yet to intend is to be conscious of intending; how could I lie to myself if I am conscious of my intention to deceive myself? The problem must be modified somewhat in light of the previous considerations: the intentional project of bad faith is now a project of believing something which is by its very nature a 'matter of faith' and which if reflected on will be revealed as mere opinion. Sartre's claim is that bad faith *exploits* the nature of faith: bad faith 'is resigned in advance to not being fulfilled by this evidence ... It stands forth in the firm resolution not to demand too much, to count itself satisfied when it is barely persuaded, to force itself in decisions to adhere to uncertain truths' (Sartre 1958, 68). The intention in question, then, is not exactly the intention to deceive oneself, but the intention to set the standards of evidence low, with one eye on the fact that faith does not in any case admit of *persuasive* evidence. 'The original project of bad faith is a decision in bad faith on the nature of faith' (Sartre 1958, 68).

The rationalist's 'demand' for reasons for trust, something that it is ultimately based or grounded on *seems* to be the very opposite of this: it seems to set the standards of evidence high—or tells itself that it does. Paradoxically of course, if this demand could be met, trust is not, in fact, required. Setting the standards for evidence high, demanding reasons, can be an attempt to transform trust into reliance on objective or independent evidence, and do away with the need for trust with all its uncertainties. Computational and information technologies do apparently lend themselves to providing formal, standardised or quantified reasons which seem to operate as 'objective' or 'independent' evidence: audit trails, identification requirements, and the ability to institute systems which apparently do not introduce interpersonal 'bias'. Through these means, we *seem* to obtain reasons for trust that are external (1) to the act of trust and (2) to the value of the thing trusted for me. However, this is to be in bad faith for two reasons: first, because the very structure of trust is denied, and second because, in virtue of the structure of trust, trying to establish conclusive evidence precisely distracts attention from the behavioural and interactive efforts that must be made in order actually to support trust. The apparent setting of the standards for evidence high, in one sense (external independent evidence), sets the standards for evidence low in that it distracts from the need for one's acts of trust to be supported in a different way which does not try to abstract away from the complexity of internal relations. And no wonder, for complexity can appear simply to be messiness, from which we can hardly be blamed for wanting to extricate ourselves! To turn, once again, to the simple example of ratings of alternate booksellers on the Amazon site: the quantification of evidence distracts attention from the entire sub-structure of values, motivations, reasons, evaluations, items and processes valued, etc., by which the ratings are supported, and which one is having to operate within anyway, but with all the complexity neatly packaged in a number—or a formula: 95% positive ratings out of 1,850 ratings. A trivial example, perhaps, but the rating system and the quantification of trust which are facilitated by digital systems is becoming increasingly prominent in the organisation of social life.

A different example is construing trust as security. To ask 'When is a system secure?' is another way of asking 'When can we trust it?' and in fact trust and security are often dealt with as though they are just the same thing in the computer science literature on the topic. Computer scientists (quite rightly) devote themselves to researching how to ensure security, but very often, the 'weak link' is seen as the user. Thus, it is bemoaned that computer scientists can design the most sophisticated encryption possible, but it just takes one user to leave a CD on a train, or lose a laptop, and all their efforts to secure the privacy and confidentiality of personal data contained in digital data sets come to nought. Here, something has gone wrong with the initial framing of trust—as something which is secured technologically, computationally, and independently of, separately from human interaction (though not in the absence of it, as the HCI literature on trust and security attests). In other words, the systems for establishing security and the security of that which they wish to establish (human identities, human purposes and motivations) are seen as external to each other. On the one hand, there are very high standards that must be attained for the system to be considered secure and trustworthy—all this is eminently rational. On the other hand—and by the very same token—there are low standards with respect to understanding the human, technological, digital and material interlacing and intertwining of trust.[6] But why does this deserve the epithet of 'bad faith', and why—if I am right—is it not simply a mistaken belief? The reason, I suggest, is that there is self-deception involved in these ways of construing trust as security, and that, moreover, it is a form of self-deception right at the heart of our relationship with technology. When some form of technologised rationality is at work, we can pretend that we are not subject to the seeming unpredictability and randomness of human-technological complexes (the equivalent of human freedom for Sartre's waiter). To call this 'demand for evidence' a form of bad faith or self-deception implies seeing it as a kind of moral failure. In the next section, we shall see why this might be the case.

Moral aspects of trust

Either trust always does have a moral dimension or, for those who claim it is only a matter of epistemic belief and not moral action and can therefore include trusting objects as well as trusting other agents, there is a type of trust which is specifically geared towards other human beings. As Annette Baier (1986), and many others, have pointed out, in this form of trust the truster cannot specify in advance what exactly the trusted person is to be trusted for, since this undermines trust. Thus for example, if I trust my babysitter, I cannot specify each and every thing that I expect him to do or not do in my absence. To do so is not only impossible on a practical level (for, how could I foresee every circumstance for which trust may be required) but my attempting to do so would show that I do not, in fact, trust my babysitter. Rather I mistrust him, since I do not believe that he will be able to use his judgement to do what is necessary in an unforeseen situation. To trust someone is precisely to underspecify what they are trusted to do or not to do. By the virtuous circularity of trust, this form of trust is also *more likely* to result in people behaving in a trustworthy manner than mistrusting them in advance. It is this strongly encoded social expectation and obligation that the act of trust institutes in normal intersubjective relations that makes possible the suspension of the demand for evidence on which trust relies, according to Mollering (2001).

However, another aspect of morality is in the tendency of trust—or a particular demand for evidence for trust—to itself be culpable of bad faith, and hence of a type of moral failure. The moral failure that is instituted against oneself is one thing (not negligible, but I will not attempt to broach it here); but the moral failure of attempting to pin down those others whom we encounter in our Internet transactions to a standard according to which they cannot but fail is a failure which is more troubling. As the negation of the virtuous circle mentioned above, this vicious circle's ultimate result, if it were to be left untrammelled, is to undermine trust in society: the more that we insist on technologised rationality in so-called trusted computational systems, the more ordinary users are seen as the ones who fail the system; the more audit trails of the type that digitised systems are so good at constructing as evidence for our 'trust', the less we are able to trust others (teachers, social workers, health care professionals, and any others) in our social worlds, and the less trust, the less well-being.

However, there is a counterbalance to this picture. There are many different kinds of digital and virtualised environments, with different framings of trust—and often with several at work at the same time. Virtual Research Environments[7] are good examples. These are Internet-enabled environments through which researchers can access and share data and other resources (for example, documents, or software tools for analysis of data, or high-performance computing resources) for their research. These function like an online distributed workplace. In constructing these environments and getting scientists to use them, it soon became very clear that trust is an essential ingredient without which the entire enterprise could not succeed. Data is a treasured resource: why would scientists share it with others when they do not know how it will be used by those others, whether they will be recognised and acknowledge as the 'owners' of the data, or whether they will be scooped? Similarly, why should they trust the data deposited by others? The questions are endless. Clearly, trust cannot be a linear process in these environments: one cannot first demand evidence and then trust—or nothing will be deposited, and once deposited it will not be used. The way in which it actually works is by means of a kind of bootstrapping from other trusted environments—for example, getting an environment started with groups of scientists who already know each other in other contexts.

How do items actually come to be trusted in so-called virtual environments? In some virtual research environments, scientists share workflows—that is, the step-by-step procedures for conducting experiments, according to a standardised format and notation for such experiments in the relevant discipline[8]. In this scenario, there is a great deal of investment of time and resources into undertaking an experiment. The benefits of sharing workflows are potentially great. In particular, for individual researchers it could greatly shorten the time that they spend finding other relevant experiments, building on them or modifying them slightly, possibly replicating them and so on. It is a situation which is 'redolent' of all the conflicting values of scientific research: the value of sharing, but also the value of competition and getting to the publication first, the value of the workflow in itself, the interest which it raises, the tremendous value of process and instruments for the scientific process, and so on. In order to work in that domain, trust is essential. But how to trust and what to trust? It is possible that in the environment which an individual scientist accesses online nothing much does emerge as a reason for trust. There are workflows, and perhaps some of them have been downloaded by others, but how does one know whether to trust them factually (are they correct? accurate? deliberate lures?). How to know whether they are *even interesting as science*? The environment in a technical sense alone does not offer up much by way of giving a reason for trust. No particular workflow succeeds in actually being 'cathected' with value, they are not picked up by others, and people do not deposit their own. They are like the wheel lying on its side in the list of examples given by Merleau-Ponty and quoted above (p. 106). But once these workflows become part of an interactive site where workflows are not standalone objects, but contextualised within social and other relations, the workflows are experienced differently. Recall that Merleau-Ponty remarks that the difference between the wheel on its side, and the wheel bearing a load is different *for sight*. In this context, items like workflows which are richly interconnected with others 'come alive', become usable. For example, who has deposited the workflow, who else has used the workflow and recommends it, the comments and suggestions in informal language which transmit know-how that cannot be captured in the formal notation, and so on.[9] This computational system works on a different potentiality of current Internet technologies, that is, their ability to create networks of relations which people experience as continuous with their social lives. Once the experiments are set in this way in a broader discourse, the virtual environment is connected up with the physical laboratory and its sphere of activities, and there is an important shift. No longer inert, dead objects, the workflows gain their significance and trustability from being embedded within a context which makes it possible to interact with them, and, most importantly, to interact with-others-with-them. This is crucial for the moral aspects of trust. By allowing for a high degree of interconnectivity between people, objects, and processes, within domains which are defined around common goals and pursuits, these environments allow epistemic and other moral values and virtues to play themselves out around the objects exchanged, rather than isolating them. In their bare forms, the workflows do not have significance, in the dual sense of value and meaning, for any isolated potential user; they gain in significance as they become embedded within a web of interactions that crucially involve and inspire trust, e.g., wiki postings, acknowledgement of failure, apparently well-meaning efforts to help others facing similar problems, etc. At the same time as they gain significance, the question of what might be a reason to trust them is also answered: it is the whole system of interactions around them, interpersonal, discursive, physical and technological. In these scenarios trust is tightly coupled to actions and interactions in which the moral aspects are not directly targeted, but indirectly given scope to be enacted.

The virtual / physical interworld

Another distinction which is not useful for an understanding of trust on the Internet is that between virtual and physical objects. By their very nature, digital objects 1) can be replicated extremely easily; and 2) can be manipulated almost as easily. Add to this the multiple modes of transmission made available by the Internet, and we have a world of proliferating objects, but objects which also change (for example, sequences are changed, images are cropped, attributions deleted, colours saturated or changed, and so on). This multiplicity of indiscernibly different objects creates an anxiety: how to choose, how to value any one (or any one set) of these objects? Very often, this is manifested in the form of not knowing which of these digital or virtual objects actually represents the real, physical object. We want our objects tethered to some anchor, so that we can be sure that they are what they are. In many digital environments, the anxiety of provenance is the way in which this is given expression, as illustrated by this passage from a National Science Foundation site:

> Provenance refers to the knowledge that enables a piece of data to be interpreted correctly. It is the essential ingredient that ensures that users of data (for whom the data may or may not have been originally intended) understand the background of the data. This includes elements such as, who (person) or what (process) created the data, where it came from, how it was transformed, the assumptions made in generating it, and the processes used to modify it.[10]

Very often, one of the ways in which the discourse on trust in virtual environments is framed relies on a distinction between virtual (or digital) and real. It culminates in a Platonic anxiety about the relation of a representation to what it represents.[11] This also means that the virtual/digital is seen *as a representation*. Evidence for whether we can trust these objects or artefacts is then taken to be dependent on their relationship to the real thing: whether that relationship is in some sense a *veridical* one. Walter Benjamin's 'Art in the age of mechanical reproduction' (1999) puts reproducibility in a different perspective. For example, Benjamin remarks on the very different positioning of the camera relative to actors and scene in film as opposed to in theatre, and speaks of the fact that there are not two spaces in filming, but rather a 'thoroughgoing permeation of reality with mechanical equipment' (1999, 227). In just the same way, our physical spaces are thoroughly permeated by the digital, in this age of digital cameras, mobile phones, smart cards, and pervasive computing.

Flick'r is a site which would, if it could, creak under the weight of the millions of images it hosts. Users of the site upload their images to be shared with select others, or with the public at large. Browsing through the site, I came across an image taken in an art gallery. The image is of a man taking a photograph of a painting in the gallery. This is now a familiar sight. We have all seen it occurring and we may indeed have done it ourselves (other prohibitions in the galleries allowing!).

This is (among other things) a self-referential picture, a picture of a person taking a picture, both of which very possibly would turn up on Flick'r. Apart from highlighting the 'permeation of reality with mechanical equipment' of which Benjamin writes, the picture nicely captures the interconnections and crossing over between physical and digital objects, real and virtual spaces, offline and online worlds. We are no longer dealing with distinct worlds but with virtual-physical interworlds.[12] The reason why the picture taken by this person may have any value is not because it represents the painting, or because it represents someone taking a picture of a painting, but because it is an action, and interaction with the painting in the gallery space, with others in the gallery space, an interaction which is already geared towards the digital archive (such as Flick'r itself), and further activities, such as gathering, collecting, and hoarding; tagging, manipulating and embedding in mashups; exchanging, sharing and distributing.

Trusting these digital artefacts is not a matter of trusting them as a proxy of some real object. Rather, trusting them is a matter of seeing them as embedded in possible actions and interactions, which cross virtual and physical environments, in terms of the values held or goals pursued. Their reality does not depend on whether they are rendered physically or virtually, but on whether they fit into networks of actions and interactions, and what it is possible to do, achieve or accomplish with them.

Conclusion

The discussion in the previous sections tries to deconstruct the distinctions that often inform discourse around trust in the Internet. Starting from the position that to demand reasons for trust from outside a system of internally related reasons, values and enactments of trust is like being an epistemological sceptic: such a stance is ultimately not just idle, but senseless. It is only from within such a system that a reason for trust can emerge. This account does mean that there is a certain circularity in trusting and having reasons for trust, which is not, however, a pernicious circularity. Importantly, it does not automatically preclude mistrust, or mean that trust is never misplaced. But in this, these Internet-enabled environments are no different from the rest of our social lives.

(Virtual) entities, of whatever type (data, audio, visual and textual content, etc.) do not have value in and of themselves; (virtual) entities do not have signs of trustworthiness, reasons to be trusted, in and of themselves. Rather these three elements (trust, value and reasons) are interdependently constituted in the virtual/physical interworld in which objects flow between different kinds of spaces, and are available for different kinds of actions and activities, and in which there is a high degree of interaction, mutuality and reciprocity. The demand for reasons or evidence cannot be fulfilled except from within this set of internally related reasons, values and enactments of trust.

How then to meet the challenge of trust in online technologies in terms of these reflections on trust? First, by being wary of abstractions around trust. In some cases, abstractions will always be necessary, but it is also necessary to understand that abstraction is non-neutral and changes the way in which trust functions. If we recognise this in advance, we are more likely to be prepared for consequences of the disturbance or disruption that the abstraction will undoubtedly bring, and even factor it in from the outset and actively design the human-computational system as a live evolving one in which one affects the other.

Second, by being wary of technologised rationality—which is, after all, just another form of abstraction, or counting as eminently rational that which comes naturally only to the technologies. Third, by creating environemnts that affirm the intricately inter-related values, reasons, evidence, actions, objects, processes and people that actually support trust. And finally, perhaps paradoxically, by shifting attention from trust as an overarching abstract and grand category, to the other more modest, apparently less significant attitudes, values, motivations, transactions, purposes, and so on, which support it. By attending to these rather than to trust per se, we could avoid being blinded by the grand challenges of trust, and succeed anyway in moving towards environments where there is, in fact, more trust and even more justified trust.

Acknowledgments

This paper has benefitted from discussions with many people, including other participants at the Workshop on the Philosophy of Virtuality, organised by the editors of this book. I would like to thank Charles Ess for his patient and careful reading and editing, Katherine Morris for discussion on Sartre, and especially Giovanni De Grandis for discussions on this topic, and so many others. Any mistakes or limitations of this article are of course entirely due to me.

Notes

1. In this paper I do not directly tackle the question of trusting other people online, although, as I hope will be obvious in the paper, this is implied throughout.
2. Ultimately, the demands of trust that it be well-placed can be made only from within an appropriate intersubjectively shared world. Elsewhere (Carusi, 2008, 2009) I have argued that this intersubjectively shared world demands a common perceptual as well as conceptual system; that there is an ineliminable moral component to trust, and that for such a world to be established requires trust in advance of reasons. Without advance trust, nothing can count as a reason for trust; nothing can count as a candidate for trust. Trust without the appropriate positioning in the intersubjective space is radically misplaced.
3. For the distinction between external and internal relations, I am greatly indebted to Katherine Morris, in conversation as well as through Morris 2008, especially pp 43-6 and 79-81.
4. See for example the LiquidPub project: http://liquidpub.org/.
5. See Mollering 2008: However, research also shows that actors are "poor deception-detectors and yet are overconfident of their abilities to detect deception" (Croson 2005: 113; see also Ekman 1996, quoted in Mollering 2008:12).
6. The vocabulary is derived from Merleau-Ponty (1968)—as are many of the associated ideas.
7. Also called collaboratories, cyberinfrastructures or e-infrastructures. See Carusi & Reimer (2010).
8. This discussion is based on the Virtual Research Environment myExperiment: www.myexperiment.org.
9. A more detailed discussion of this environment can be found in De Roure, Goble & Stevens (2009).
10. www.nsf.gov/awardsearch/showAward.do?AwardNumber=0455993. Accessed [date to be inserted]
11. An important point of comparison is with the treatment of writing in the *Phaedrus* (See Plato, 1998)
12. I have borrowed this term from Merleau-Ponty (1968).

References

Baier, A. (1986). Trust and Anti-Trust. *Ethics*, *96*, 231–260

Benjamin, W. (1999). Art in the Age of Mechanical Reproduction. In *Illuminations* (211–244). London: Pimlico.

Carusi, A. & Reimer, T. (2010). *Virtual Research Environment Collaborative Landscape Study*, JISC. Available online: http://www.jisc.ac.uk/media/documents/publications/vrelandscapereport.pdf

Carusi, A. (2009). Implicit Trust in the Space of Reasons and Implications for Design. *Journal of Social Epistemology*, *23* (1), 25–43.

Carusi, A. (2008). Scientific Visualisations and Aesthetic Grounds for Trust. *Ethics and Information Technology*, *10* (4), 243–254.

Croson, R.T.A. (2005). Deception in Economics Experiments. In C. Gerschlager (Ed.), *Deception in Markets: An Economic Analysis* (pp. 113–130). Basingstoke: Palgrave Macmillan.

De Roure, D., Goble, C., Stevens, R. (2009). The Design and Realisation of the myExperiment: Virtual Research Environment for the Social Sharing of Work Flows. *Future Generation Computer Systems, 25*, 561–567.

Ekman, P. (1996). Why Don't We Catch Liars? In: *Social Research, 63*, 801–817.

Hume, D. (1902). *Enquiries Concerning Human Understanding and Concerning the Principles of Morals*, (ed.) L.A. Selby-Bigge. Oxford: Clarendon Press. [Originally published 1777.]

Gambetta, D. (1994). *Trust*. Cambridge: Cambridge University Press.

Goldman, A. (1999). *Knowledge in a Social World*. Oxford: Oxford University Press.

Merleau-Ponty, M. (1962). *Phenomenology of Perception*. Translated by C. Smith. London: Routledge & Kegan Paul. [French original 1945]

Merleau-Ponty, M. (1968). *Visible and Invisible*. Translated by A. Lingis. Evanston, IL: Northwestern University Press. [French original 1964.]

Mollering, G. (2001). The Nature of Trust: From Georg Simmel to a Theory of Expectation, Interpretation and Suspension. *Sociology, 35* (2), 403–420.

Mollering, G. (2008.) *Inviting or Avoiding Deception through Trust: Conceptual Exploration of an Ambivalent Relationshiop*, MPIfG Working Paper 08/1. Cologne: Max Planck Institute for the Study of Societies.

Morris, K.J. (2008). *Sartre*. Oxford: Blackwell.

Pettit, P. (1995). The Cunning of Trust. *Philosophy and Public Affairs, 24* (3), 202–25.

Plato. (1998). *Phaedrus*. Translated and edited by James H. Nichols Jr. Ithaca, NY and London: Cornell University Press.

Riegelsberger, J. Sasse, A., and McCarthy, J. (2005). The Mechanics of Trust: A Framework for Research and Design. *International Journal of Human-Computer Studies, 62*, 381–422.

Sartre, Jean-Paul. (1958). *Being and Nothingness*. Translated by Hazel Barnes. London: Routledge. [French original, 1943.]

Simmel, G. (1950). *The Sociology of Georg Simmel*. Transl., ed. and intr. by K. H. Wolff. New York: Free Press. [German original, 1908.]

Simmel, G. (1990). *The Philosophy of Money*, 2nd edn. London:Routledge. [German original, 1900.]

Strawson, Peter. (1985). *Scepticism and Naturalism. Some Varieties*. London: Methuen.

CHAPTER SEVEN

Trust, Lies and Virtuality

BJØRN MYSKJA

Introduction

Trust is essential for a well-functioning human life. We are social beings who depend on each other for survival and for living meaningful lives, and if we did not establish trust-based relationships with others, our lives would be miserable. Some sociologists even argue that trust plays a special role today, as the modern human life, characterised by technology-driven welfare, is also a condition of increasing complexity and new patterns of risk due to the same technology (Beck, 1992, p. 21). For Niklas Luhmann (1989), trust is a human way of handling the complexities and risks of modern society, which include trust in institutions and technologies.

The most fundamental form of trust is not a novelty caused by relatively recent societal and technological changes. Trust and distrust have always been part of the human condition, as can be read in a variety of sources such as the Norse poem *Hávamál*, the Bible or Aristotle's *Nicomachean Ethics*. In many cases, it is treated in terms of friendship and enmity, but the issue is what we now call trust. Interpersonal trust is one between people who have been close over a long period of time, but also the kind between strangers who meet for one day on the road, expecting never to see each other again.

I wish to discuss some of the challenges to this kind of interpersonal trust posed by new modes of communication within what is called virtual reality. In the last decade or two, increasing parts of our modern lives take place online, in different kinds of social forums. In this context I am not interested in online communication as part of commercial or academic activities, but as a forum for social interchange. The interesting issue is to what extent a communication medium with a tradition of artificial environments, fictitious characters and restricted opportunities for distinguishing fact from fiction can be a basis for interpersonal trust. One could say that the online world and its inhabitants are more or less unreal or not true, while trust and trustworthiness seem to be strongly connected to truthfulness. The problem is not only that it is difficult to check whether my online friends speak the truth without engaging with them offline, but that lies or misrepresentations are integral elements of the social interactions in parts of virtual reality.

My point of departure is a discussion of the connections between trust and veracity. Although a trustor does not always expect her trustee to tell the truth, she expects truthful behaviour. This is brought out in the Kantian understanding of morality as autonomy, where trust and veracity are central elements of moral behaviour. However, Kant's ideal theory is moderated in his later, more practical, empirical ethics, where we have to deal with a social reality of imperfect human beings. Here it is morally acceptable to distrust oneself and others, and even to adjust the demand of truthfulness in order to work for moral improvement. An important condition is that we do not deceive ourselves and others by this deceptive behaviour: rather, we reciprocally recognise that we engage in this kind of deception.

There is an element of fiction in the Kantian philosophy generally, which has been called his philosophy of the "as if." This kind of fiction is not one of misleading others, but a way to understand or relate to the world through hypothetical thinking. In this way it is a kind of deception as we think or act *as if* reality were different. To begin with, this is a benign deception that is crucial for art and aesthetics. We regard the piece of art *as if* it were nature, while knowing that it is not. Online chat rooms seem to share both kinds of Kantian benign deception: the chat room world is one of art, and the people inhabiting it reciprocally recognise that they may deceive each other. I argue that this is not morally problematic in itself, but is to some extent an obstacle to trust because we cannot know where deception ends and trust can begin. Although interactions online seldom put us at risk until they move offline, it is increasingly acknowledged that people do develop valuable relationships online. Therefore, issues of trust and deceived trust may be increasingly important also for this kind of social interaction.

Trust and veracity

There is an internal relation between trust and truthfulness. When we trust someone we usually assume or require that they speak the truth. Admittedly, not all instances of trust involve assumptions of veracity; often trust is directed at the competence of a person or institution. When I trust my plumber, I trust that he knows how to fix my leaking faucets, and I do not care about his general honesty. Truth may become an issue related to trusting his promises regarding timing of the work or when we discuss payment, but not directly when it comes to trusting his competence. But even when trusting competence, truthfulness seems to play a role. We usually trust someone's competence based on their own explicit or implicit claims about their abilities. Now, my trusted plumber may speak truthfully but be mistaken when he says he possesses the required know how. Then it is only my trust in his competence and his own assessment that is broken. But in most cases of trust, even when competence is the issue, truthfulness plays a crucial role. In this essay, my concern will not be with trust in competence, but with the role of trust in relations where issues of truthfulness may be essential, and the more special cases of internet relations.

Net interactions are often described as taking place in a "virtual reality" which should be distinguished from the "real" reality. The main difference between these two modes of intersubjective contact seems to be that one is in the physical presence of others, whereas the other is electronically transmitted. However, this does not really account for the virtuality of virtual reality. Telephone conversations are also electronically transmitted without physical proximity. More important for the distinction is the partial fictionality implied by calling the reality "virtual."

This relationship between trust and truth is captured by Sissela Bok (1989), who says that veracity is fundamental in trust-based human relationships:

> I can have different kinds of trust: that you will treat me fairly, that you will have my interest at heart, that you will do me no harm. But if I do not trust your word, can I have genuine trust in the first three? (p. 31)

At first glance, this rhetorical question seems to capture an important insight into human relations, but it is probably not accurate in an unqualified way. Although truthfulness seems to be an essential aspect of many instances of trust, it is not a necessary element. It is deeds, not words, that count in human relationships, and we may very well trust someone based on their previous acts, even if we know them to be less than truthful in speaking. But then we, metaphorically, trust their honest character rather than their words, so in a sense we can say that truth is still essential. In many cases people say the opposite of what they mean for fun, expecting others to understand what they really mean. Their honesty is trusted even if the literal meaning of their words is not believed. Sometimes we do not know what people intend to do, but we trust that they will do the right thing even if they express something different verbally.

A related but different case may be when we trust someone despite their expressed intention to not take care of the interests we want them to. The reluctant heroes such as *Casablanca*'s Rick Blaine or *Star Wars*' Han Solo refuse to help out, but the people in need still count on them. And in the end, the heroes act contrary to their words, and the trustors can trust them to not keep their word. Several reasons for this paradoxical behaviour can be suggested. (1) The trustee is merely pretending to shy away from the responsibilities of trust, and he knows that the trustors know this. The refusal is mere pretence, recognised as such by both parties. (2) The trustors perceive the true nature of the trustee, who himself does not realise his own underlying character as trustworthy and reliable. (3) The trustee has a true change of heart, caused by reflection of the moral challenge they initially rejected. As has been pointed out by many, placing trust in someone may contribute to make them behave in a trustworthy way (Elster, 2007, pp. 350 ff.). In this respect, trust can function as a moral challenge.

Also, even if perceived truthfulness were necessary for trust, it would not follow that general transparency is a necessary requirement. Even if trust depended on the truth of the words we utter, we would not have to tell everything. Some essential cases of trust seem to require that the trustor does not get full information. The expressions "I trust you!" and "Trust me!" implicitly tell that the trustor need not know all the details in the matter. This is connected to paternalistic elements in trust, exemplified by Humphrey Bogart's Rick Blaine in the movie *Casablanca*, saying to Ingrid Bergman's Ilsa Lund: "Let me do the thinking for both of us," implying that she does not need to know his plans or the way he intends to fulfil them; she only needs to trust him. This is similar to aspects of religious faith, where the ultimate version is that faith in God requires that His existence cannot be proven and that religious belief is non-rational (Kierkegaard, 1994. pp. 44 ff.).

Usually when trusting someone, we do not expect them to keep us informed continuously about what they mean and think. So even if there seems to be a connection between trustworthiness and truthfulness, the connection is not a direct one. But we expect those we trust to answer us truthfully if we ask them. This is not unproblematic, though. In many trust-based relations, asking for this kind of information may be taken as a lack of trust. John Weckert (2005, p. 98) has pointed out how being trusted affects our self-esteem and contributes to our autonomy, and how systematic distrust undermines autonomy. It is impossible to choose freely to do the right thing if you are controlled and continuously have to account for your work. But not all kinds of checks and controls undermine autonomy. The fact that there exists institutionalised distrust ensures that we can trust the transportation systems, the health services and other central aspects of modern life (Grimen, 2009, pp. 90 ff.). As Luhmann (1989) famously has pointed out, this aspect of trust reduces the complexities of living in a modern society. Thus, there is a clear but complex relation between trust, distrust and autonomy (Grimen, 2009, p. 101).

Trust and autonomy

The relation between trust and autonomy is brought out clearly in Onora O'Neill's (2002) analysis of the Kantian approach to trust. The core of morality is expressed in the Categorical Imperative which in its first formula says: "Act only according to that maxim whereby you can at the same time will that it should become a universal law." (Kant, 1965, p. 42) Autonomy means self-legislation and expresses the idea that only by following rules that we ourselves have given can we be truly free. Moral choice is expressive of human rationality which, on O'Neill's interpretation, is our communicative and reason-giving personality. It follows that Kantian autonomy includes the ability to communicate the reasons for our actions to others. Actions that are harmful to others are contrary to this requirement, which means that any kind of deception or coercion of others is ruled out. O'Neill argues that this is exemplary of trustworthy actions (2002, p. 97). We trust those who do not deceive or coerce us. So, if we accept her interpretation of Kant, the essence of morality as expressed in the Categorical Imperative is to act in a trustworthy way.

This conclusion is supported by the second formulation of the Categorical Imperative: "Act in such a way that you treat humanity, whether in your own person or in the person of any other, always at the same time as an end and never merely as a means to an end" (Kant, 1965, p. 52). If we deceive others, we do not treat them as ends, since we deny them the opportunity of freely choosing how to act. This is impossible if they are wilfully misled. Unless we act according to the Categorical Imperative, which includes being truthful, we are not trustworthy and there is no sound foundation for trust. Still, Kant holds that the main vice is not harm to others or oneself but that by lying, a human being "throws away, and, as it were, annihilates his [or her] dignity as a human being." (Kant, 1996, p. 429) He goes on to say that one makes oneself less than a thing, as a thing can be useful, but lying as communication of "the contrary to what the speaker thinks on the subject is an end that is directly opposed to the natural purposiveness of the speaker's capacity to communicate his thoughts" (ibid.). Harm is less of a moral problem than lying, as the harmed person, be it myself or others, still retains his or her humanity. There is an element of "crime against nature" in this interpretation. But as I pointed out initially, there are many cases where we communicate through deeds rather than words, and where the lies cannot meaningfully be classified as counter-purposive in the way Kant does in the passage cited above. We have a wide range of ways to communicate our meanings and intentions, many of them not identical with the literal meaning of our words. The issue must be to what extent the audience understands what we mean rather than the literal meaning of the words we use. It is also worth noting that when Kant discusses examples to illustrate the moral problem with lying, the recurrent theme is that by attempts at morally motivated lying, we implicitly take responsibility for that which we cannot control, namely the consequences of our acts (Kant, 1996, p. 431; Kant, 1968, p. 639; Neiman, 2002, pp. 73 f).

A reconstructed Kantian account, following the interpretation of O'Neill, regards the core problem with lying to be that it is an act of deception and a failure to act in a trustworthy way. We should avoid Kant's absolute demand of truthfulness, which is contrary to some of our basic moral intuitions. If we regard his examples of lies, it is clear that the practical moral problems are primarily connected to harm in the form of deception of others. There are sound reasons for modifying this part of his account (Korsgaard, 1996, pp. 133 ff.) by allowing for lies that do not deceive our communication partners, as long as we make clear that this does not weaken the strong connection between truthfulness and trust. It is our basic moral duty according to the Kantian moral philosophy to act in accordance with the trust others place in us, that is, to act in a way that makes us deserving of trust. We can be trustful in this sense, while allowing for some evasions of telling the truth.

I have argued earlier (Myskja, 2008) that Kantian autonomy also means that we have a duty to trust other people. Distrust cannot be universalised, because trust is a basic condition of social life. Even if we choose to live in solitude our whole grown-up life, we cannot become adults without a cooperative society. Such a society would not function without trusting human interaction. This conclusion is strengthened when considering the second formulation of the Categorical Imperative. When we distrust others, we assume that they do not act according to the Moral Law. But it is only when acting according to the Moral Law as expressed in the Categorical Imperative that a person actually is an end in itself. All other motives of choice are heteronomous, as they are not expressions of self-legislation. By distrusting someone, we assume that he or she is motivated by non-moral incentives. Through our distrust we fail to treat the other as an end in itself.

Non-ideal ethics, truth and trust

We have seen that Kant's demand of absolute truthfulness is problematic in regard to our everyday communication. Several authors have pointed out that there is a distinction between Kant's "pure" *a priori* and his "impure" empirical ethics (Louden, 2000, pp. 5 ff.), which is evident in the area of veracity. In his empirical writings and lectures, Kant seems to accept that if we treat others as always acting morally we easily become victims to, or even tools of, evil. He is recorded as saying that "as men are malicious, it cannot be denied that to be punctiliously truthful is often dangerous" (Kant, 1997, p. 228). This kind of truthfulness is not only potentially harmful to ourselves, but may also harm others. Clearly the cases of lying to protect an innocent from pursuers with evil intent or lying to a child about the quality of her paintings or piano playing to encourage her to continue practicing do protect people from avoidable harm. And more importantly, I am in a position to prevent that harm.

In such cases we seem to be in situations where our duty to avoid harm to others conflicts with our duty to be truthful. Kant, however, says that duties cannot conflict, only *grounds* of duty may conflict (Kant, 1996, p. 224) Although we seem to have two conflicting duties, this is mere appearances. There is a logical reason for this: An overriding principle in morality, according to Kant, is that ought implies can. Since we cannot simultaneously fulfil two conflicting duties, it follows that they cannot both be duties. They are merely *grounds* of duties, and as soon as we realise that we have one duty—to avoid harming a fellow human being, we have no duty to be truthful—or vice versa. We have one duty and one ground of duty which does not lead to a duty in this case (Herman, 1996, p. 79 f.).

This solution to the problem of conflicting duties has at least two problems. The first is that it is counterintuitive. Most of us would experience the conflict as a real conflict even if we have decided to act on one of them. In this case it would perhaps be most obvious if we have chosen the duty of truthfulness and we then become tools of the wrongdoer's harm to a fellow human being. We would feel regret and shame. Now Kant argues that sympathy with suffering that we cannot alleviate just increases suffering in the world and should be avoided on moral grounds (Kant, 1996, p. 457). He would probably say the same about regret and shame in this case, but that would not save him. We should not accept a Kantian rejection of emotions that are morally essential for our self-understanding, because we share responsibility for a given evil as long as we could have acted otherwise.

This is the second problem with Kant's solution to the problem of conflicting duties. As long as there is no principle for choosing between the grounds of obligation, we cannot say that we could not have acted otherwise. We could have chosen the other ground of obligation, and the choice is apparently arbitrary. Kant has only moved the problem of conflicting duties one step, to their grounds, a solution that is merely one of words. There is a conflict between grounds of obligations, and we must choose one of them, without any hope of escaping the regret of not choosing the other—even if Kant says that we are not to blame. It is also a fact that his theory leaves room for choosing lies in certain situations, even if that is prohibited in his ideal theory. It is also possible that there is some suggestion in his non-ideal theory for rules of thumb for choosing when the ground of the duty of veracity is overruled.

Kant is a keen observer of human behaviour and entertains no illusions that people are morally good. Even if we are good, we have no way of knowing that we are, since action in accordance with the moral law is not necessarily action motivated by the same law. Kant believes that we often act morally merely to appear better than we really are, and that this is a form of deceiving others. We may appear to have a trustworthy character, but the evidence for such a character are merely acts that could have been expressive of a trustworthy disposition: but it is more likely that these actions are based on a deceptive disposition. This is a result of our social and imperfect natures. We crave social belonging and acceptance, but we are not morally good in a way that would secure this. So, in order to achieve the first on the basis of the second, we must deceive others, pretending to be better than we really are. The interesting point in this somewhat dismal picture is that Kant says that people are not fooled by these deceptions. We expect each other to pretend to be better than we really are (Kant, 1983, p. 67).

We can interpret this in two ways. We can see it as a description of a human state of reciprocal deception and distrust. We cannot and should not trust each other. It is the state that is imagined when we use the first formulation of the Categorical Imperative as a test of the acceptability of lying: could I will a world where the general rule is that everybody deceives each other? Actually, I live in such a world whether I will it or not. The regrettable fact is that those who do not participate in this game of pretence will appear as morally worse than the rest, despite the fact that they are better, due to their honesty. We can, however, also see this as an interesting anthropological observation by Kant. We do not actually deceive each other when we all pretend and all know that we all pretend. We are engaged in a game of pretences where the reciprocal recognition of deception is part of what binds us together. The underlying condition here is of course that we do not engage in this game of pretences in order to take advantage of others, to harm them or to make them our tools. It is part of human sociality, especially, as Kant says, in civilised society because we want to appear as good human beings.

The paradox is that without veracity "social intercourse ceases to be of any value" and the "liar destroys this fellowship" among men according to Kant (1997, pp. 200 f.), while he is also reported as saying:

> Man holds back in regard to his weaknesses and transgressions, and can also pretend and adopt an appearance. The proclivity for reserve and concealment rests on this, that providence has willed that man should not be wholly open, since he is full of iniquity; because we have so many characteristics and tendencies that are objectionable to others, we would be liable to appear before them in a foolish and hateful light. But the result, in that case, might be this, that people would tend to grow accustomed to such bad points, since they would see the same in everyone. (1997, p. 201)

That is, when it is recognised as part of a social game we all play, it is morally acceptable to pretend to be better than one is under non-ideal circumstances. If we did not pretend, we would get used to immoral behaviour, thinking that this is how it should be. This indicates another paradoxical feature of Kant's view on lying. This kind of benign, socially recognised deception is not only morally acceptable but may even foster moral improvement. In an echo of Aristotle's idea of becoming what you do, Kant seems to think that when we act *as if* we are good (i.e., when in fact we are not, so that we are deceiving others), we may become better. Society at least is served by this pattern of make-believe. Perhaps we may become virtuous due to imitation of virtuous behaviour? This is not deception in the way an outright lie for our own benefit is, but a form of deception that appears to be necessary both for our own social preservation and for creating a morally better community. Honest revelations of our moral flaws are often more harmful than deceiving others by pretence.

Trust and the "as if"

The way we choose to present ourselves and the way others see us are two not totally connected aspects of trust. Even if you present yourself as better and more trustworthy than you are, my trust in you is partially independent of both your appearance and your real nature. In his analysis of trust, John Weckert (2005, pp. 101 f.) discusses cognitive and non-cognitive accounts, pointing out that a number of instances of trust cannot be reduced to expectations or beliefs. He suggests that trust is an "intertwining of the attitudinal and the cognitive" (2005, p. 102) and can best be captured in the notion that we see someone as being trustworthy. This "seeing as" he compares to a Kuhnian paradigm, noting that we usually do not think much about trust unless our trust is broken.

Weckert (2005, p. 112) supports this idea of trust as a "seeing as" with Wittgenstein's reflections on the significance of aspect changes, such as the duck-rabbit (Wittgenstein, 1992, pp. 193 ff.). The picture does not change when I suddenly see the rabbit rather than the duck. Likewise, when I stop trusting someone I have trusted for many years, I see him differently, although he is the same person and may have been acting the same way all the time. But seeing someone as trustworthy is of course not independent of the actual behaviour of the person or institution I trust. As Hardin (1996, pp. 28 f.) has pointed out, the crucial ethical questions discussed under the theme of trust usually concern trustworthiness. So even if trust is a seeing as, how we end up seeing someone will, in the long run, be partially dependent on how our trustee behaves. Still, trust cannot be reduced to the question of the true nature of the trustee, as a cognitive account implies. We can, as Weckert argues, choose to see someone as trustworthy or not, as long as we do not have firm evidence barring one of the options. It is well known that people differ when it comes to what they regard as sufficient evidence for trusting or distrusting someone.

Weckert (2005, pp. 113 f.) connects his idea of trust as a voluntary "seeing as" to trusting as a leap of faith where I act *as if* I trust someone, suggesting that this not only can lead to trust, but is a way to generate trust where it does not exist. We make a choice although we have insufficient evidence that it is well-founded. This is a parallel to Kant's suggestion that pretending to be good can make me good, with its Aristotelian connotations. There is, however another interesting connection between trust and a general use of "as if" thinking in Kant's philosophy that I want to explore here. Caygill (1995, pp. 86 f.) shows that the "as if" structure of Kantian arguments can be found in all three knowledge areas: theory, practice and aesthetics. The regulative ideas have a form of as-if reasoning in the theoretical realm, where we judge the world *as if* it had been created by God, and we judge humans *as if* we had a mind with a persistent personal identity. Caygill does not mention that the idea of God also has a practical implication for teleological judgements of nature, where we find particular natural laws by judging nature *as if* its laws were given by an understanding, and therefore according to the principle of purposiveness (Kant, 1987, p. 180). More important in this context is the use of "as if" in the practical realm where the third formulation of the Categorical Imperative says: "Therefore, every rational being must so act as if he were through his maxim always a legislating member in the universal kingdom of ends" (Kant, 1965, p. 438). This formulation takes account of the fact that we are social, communicative beings, striving to build a good society from the imperfect world of pretenders described above. According to the imperative, I should not pretend to be perfect, but rather act *as if* the world, including my fellow human beings and myself were perfect. When I act on this imperative, I envision a world that does not (yet) exist in order to contribute to its future realisation.

The most relevant use of the as-if structure in Kant's thinking for the present issue, is his use of it in art. Representational art, which was the background of Kant's analysis, is deceptive in the way that its quality is connected to appearing as something different from what it is. It must appear *as if* it was nature, not a human product:

> In a product of fine art we must become conscious that it is art rather than nature, and yet the purposiveness in its form must seem as free from all constraint of chosen rules as if it were a product of nature. (Kant, 1987, p. 306)

Art is a form of deception that we not only recognise and accept, but where its quality is a direct product of these deceptive qualities. If the form of the artwork appears as based on human rules, we would search for its message rather than subjecting it to aesthetic judgement and appreciation. But this deception is not only benign, but also a form of deception that is socially recognised and accepted. It is part of the convention of art that it is not what it appears to be. We could say that some of the contemporary controversies in art concern works that in different ways challenge this definition of art. The use of ready-mades ("found art") such as Duchamp's *Fountain*, a urinal displayed at an art exhibition, challenge the idea that it must appear as nature. The object is an industry product with an obvious function. Conceptual art, where the ideas of the work take precedence over its appearance, carries intentional messages undermining perceptions of it as resembling nature. These challenges, however, stimulate further exploration of the deceptive nature of art. Instead of being a work of human invention that appears as nature, *Fountain* is a work of human invention that appears as artworks usually do. Instead of being art because it appears as nature, it is art because it is non-art that appears like art. Conceptual art is challenging the "as if" by letting the artwork be governed by chosen rules. But by being transformed into art rather than merely spelling out the concepts, something beyond the concepts remains. We cannot escape the "as if" of art totally.

The Kantian account of aesthetics says that humans have a moral interest in natural beauty which we cannot have in our aesthetic appreciation of art, due to its deceptive character. We have an intellectual interest in the non-intentional purposiveness of nature that serves our moral nature, according to Kant (1987, pp. 298 f.). He uses as example someone who takes pleasure in nature's beauty, a pleasure devoid of all self-interest. This pleasure will disappear instantly if he discovers that it is not natural but artificial beauty he is admiring:

> Suppose we secretly played a trick on this lover of the beautiful, sticking in the ground artificial flowers ... and suppose he then discovered the deceit. The direct interest he took in these things would promptly vanish, though perhaps it would be replaced by a different interest, an interest of vanity ... [T]he thought that the beauty in question was produced by nature must accompany the intuition and the reflection, and the direct interest we take in that intuition is based on that thought alone. (Ibid. 299)

The knowledge that the object is fictional changes its value. We have an immediate interest in a work of art that is not present in the valuing of natural beauty, according to Kant, namely a social interest which is not necessarily moral in character. It is only an intellectual interest connected to the disinterested appreciation of beauty that is conducive of morality. This is a consequence of the contrast between the pure and the impure as well as between the real and the fictional. The fictionality of art is not morally wrong, but it cannot serve the same moral function as the aesthetic experience of reality. The question is whether this moral deficiency also affects virtual reality. When we make a fictional world that only *appears* to be real, it lacks characteristics that we appreciate for moral reasons in the real world. Does this affect the moral appreciation of the virtual realities we find online?

Virtuality, deception and trust

The virtuality of online interactions, "virtual reality," can be placed between the two forms of recognised deception discussed by Kant. Virtual reality is a form of art in the sense described by Kant. We know that it is fictional, but we must treat it as reality in order to make it work the way we intend. It is a deception where nobody is deceived—although they may be deceived on other accounts than regarding the reality of the medium. But there is an extra element of recognised deception in this virtual world. As we know that we have limited information about each other and there are many opportunities for improving on ourselves in the eyes of our communication partner, it is reasonable to assume that people present themselves as more attractive and morally better than they are in reality. There will certainly be some context-dependent exceptions, but in most cases we want others to see us as good people. This brings us to the other kind of recognised deception described by Kant.

Kant's observation about veracity and deception in human interaction takes on extra significance online. There are several reasons for this. First, most online interactions still are text-based, and we lack essential trust-fostering clues of social interactions in the offline world. Emoticons are insufficient substitutes for facial expressions and body language when we seek to interpret the intentions behind verbal communications. We find a growing literature on the significance of these essential clues to meaning and sincerity for building friendships and other trust-based relationships online (e.g., Cocking & Matthews, 2000; Nissenbaum, 2001; Weckert, 2005; Briggle, 2008; Myskja, 2008). Many hold that the lack of a large number of normal and socially established clues for interpreting the trustworthiness of our communication partners must influence the extent to which we can trust each other and build real friendships online.

It is also probable that the history of the Internet contributes to the social rules for online interaction. The early text-based MUD (Multi-User Dungeon) role playing games made room for exploring alternative identities as people did interact through fictitious identities in equally fictitious environments (Turkle, 1995). We do not have to subscribe to outdated postmodern theories of the self to accept that this sort of identity play has been central in the early stages of social interaction on the Internet. We should also assume that this background influences the implicit rules of the present online activity. MUDs have developed into contemporary graphic social interactive games (i.e., Massive Multiplayer Online Games, or MMOGs) and virtual worlds such as *Second Life* where players and participants represented by avatars interact in almost true-to-life worlds on the screen. In many online discussion forums, members are also represented by avatars, which could be taken as an indication that they to some extent play a character and therefore their words should not always be taken at face value. It serves as a warning that not everything told here is sincere or represents the considered opinions of the writer. I express this rather cautiously because many chat room interactions are sincere and people do develop valuable, meaningful, trust-based relationships where they share significant experiences of their lives. Still, these relationships are developed within what Nissenbaum calls "inscrutable contexts" which carry with them the socially accepted play with identities, always leaving room for doubt (Nissenbaum, 2001, p. 114). This doubt will only disappear totally by moving the relationship offline. Keeping this historical background for online communication in mind, we can safely assume that the room for accepted deception as regards personal character is wider online than offline.

One potential counterargument to this assumption is that today's online interaction is quite different from the conditions discussed in the earlier literature on Internet friendship and trust. Now people often use web cameras and speak together rather than write, presumably adding several of the "embodiment" elements claimed to be necessary for trust. After all, when you see the face and hear the voice of your friend, the problems listed by Nissenbaum for online trust are reduced, if not removed. That is, in addition to "inscrutable contexts," she identifies, e.g., missing identities and missing personal characteristics as obstacles to trust in interactions online (Nissenbaum, 2001, pp. 113 f.). Admittedly, the introduction of audiovisual interaction cannot fully prevent that people "cloak or obscure identity," (2001, p. 113), but that is also the case for quite a few offline interactions. Still, since several of the "bodily signals of face-to-face interaction" are present, we can no longer be tricked by a "57-year-old man posing as a 14-year-old girl" (ibid.). Nissenbaum points out that the problem of the inscrutable contexts is directly connected to the aspects that for many are seen as attractive, liberating elements of online identity play. Thus, introduction of embodiment elements will reduce the difference between online and offline interaction, improving conditions for trust but removing attractive features. Many have remarked that trust is not always morally good, such as trust between evildoers or trust in someone with bad intent. But it is also a fact that pure trust-based relationships also may lack important qualitative aspects of human relationships. Being with completely trustworthy companions all the time can be quite boring.

So even if the obstacles to trust online pointed out by Nissenbaum are reduced due to some recent changes to online interaction (to the extent these new possibilities are used), the difference with offline conditions remains due to what

Nissenbaum calls inscrutable contexts which also affect the point of missing identities. These factors, I believe, are connected to the history of online interaction, where the social interaction seems to have developed a certain subculture (perhaps more accurately, a variety of subcultures, e.g., those surrounding the norms and practices of chatrooms, virtual worlds, MMOGs, etc.) for reciprocally recognised deception. Just as our real-world interactions include reciprocally recognised deceptions, this is even more the case online where deception is even easier and there exist traditions for playing with identity within more or less inscrutable contexts. These worlds have the quality of the aesthetic "as if." We act *as if* the artificial world were a real world and the people we meet are who they present themselves as, knowing that this is not so.

This "as if" is similar to the game of pretence in real-world social interactions, but with an extra layer of pretence. In both realities I know that you are not who you pretend to be and you know that I know—and vice versa—but in virtual reality my already pretending self is taking on a fictitious identity which is mutually recognised as well. Just as Kant points out that our self-presentations of being better than we are have some morally good aspects, we can say the same about the online identity games. The chance of exploring ways of life in a fairly safe context may provide learning for offline interactions. In addition, this kind of playing with identities and contexts online, free from the constraints of our social roles may be valuable in itself as a way of richer self-expression. Because the interactions online are governed by explicit and implicit rules that are more liberal than in offline interactions, the pretences and recognised deceptions take on a different character. Deceptive behaviour that would be immoral offline is acceptable online because fictitious personalities and creative life-stories are part of the context. Not all kinds of deceptions are acceptable, but the limits are wider. Just as deceptions are acceptable in the social world offline as long as they are conducive to morality, the deceptions online are acceptable in a similar manner.

The online world is a fictitious world, a product of art. Like art, it resembles the real world, but can only function because we know it is not. We act within online worlds as if they were real, and interact with each other as if others are who they present themselves as, even when we know that this is not the case. But it would not work as an interesting and fascinating way of life unless online worlds did contain relationships and challenges that are more than mere play-acting. We do relate to people and we do trust them within the artificial frames of the virtual reality. These frames are formed by our reciprocally recognised deception: we do deceive each other, but we know that we deceive. The limits for this deception are vague, making trust seem more risky or fragile than in offline relationships. But most of our interactions in social networks on the Internet are virtually without consequences for our further life. Hardin (2006, p. 101) points out that trust is only relevant where there is risk of untrustworthy behaviour, and that relationships of trust require some contact over time. In most cases, risk becomes significant when the relationship moves offline, making trust relevant. But this is perhaps too simple. Risk comes in many forms, and also virtual relationships give room for intimacy and betrayal even when offline harm is non-existent (see, e.g. Dibble, 1993).

We can draw on Kant's analysis of the distinction between aesthetic appreciation of nature and art to see the distinctions between the moral significance of real and virtual moral relationships. Even if we are beings who orient ourselves by imagination and in many cases base our judgements according to the as-if logic, there is a fundamental moral difference between the real and the fictional. Only reality can provide the basis for significant moral judgements and provide true moral learning. Fictional choices in fictional worlds can arguably contribute to the development of our moral perception (Nussbaum, 1990, pp. 3 ff.), but as long as such choices have no impact on our lives, their moral value is limited. But one of the fascinating factors of online virtuality is that it is also reality and does have an impact on our lives as a source of joy and sorrow. Due to the blurred distinctions between the fictional and the real, we may also experience true, trust-based relationships within these fictional frames. But as the borders between the fiction and reality are obscure, even deciding when the fiction ends and reality starts is a matter of judgement. We may err and mistake the fictional for the real and vice versa.

Conclusion

There is a clear connection between trust and veracity. When I trust someone with something that is valuable for me, it is usually in a situation where they promise to act in accordance with my wishes. This promise may not be explicit—although it often is—but in most cases of trust, the trustee at least is aware of and acknowledges the trustor's wishes implicitly. If she cannot act in accordance with the trustee's wishes, the trustor expects her to say so. We can find some cases of misplaced trust where truthfulness is not required, but in normal cases, trust and veracity are parts of the same moral fundamentals of human relationships. We trust people because they are trustworthy, reliable, honest, decent people. Even if they were not, we should treat them as if they were, because that is expressive of respect for their status as autonomous, moral agents. In an ideal theory of trust between human beings, this is the whole story. Both trust and trustworthy behaviour follows from the demands of the Categorical Imperative, expressed in the formulation that we should act as lawgiving members of the kingdom of ends (Kant, 1965, p. 438)

When we take into account, however, that we are at times imperfect, deceitful, egoistic, honour-seeking social climbers—and we reciprocally recognise this state of affairs—we must adjust this moral picture. Under non-ideal circumstances it is morally justified to pretend to be better than we are, because that contributes to a better community. If I pretend to be a trustworthy person because I want to be socially respected, this is conducive to morality in two ways. First, I contribute to the social norm of trustworthiness, making this behaviour the moral norm. Second, I bind myself by my pretence. If I want to retain the image of being trustworthy, I must become or remain trustworthy. Thus the as-if of moral pretence under non-ideal conditions can stimulate the improvement of the moral community as a whole due to the strengthening of the common norms. In addition moral pretence stimulates the improvement of everyone who engages in this pretence, regardless of the effects on the community.

This picture is somewhat different under the extra layers of possible deception in social forums of the as-if world online. Although fictional worlds may provide arenas for developing moral discernment, they are also worlds with little at stake and therefore less risk. Still, the possibility of developing valuable relationships is present, and therefore also the potential for trust and deceived trust. The distinction between fiction and reality is blurred in this world, leading to a similar obscurity regarding moral rules. When deciding on moral matters within these frames, the essential issue is that we reciprocally recognise that we deceive each other regarding our identities and that is part of the attraction of this existence. The challenge is to decide when interactions turn from being deceptive to becoming sincere.

Acknowledgments

I wish to thank May Thorseth and Charles Ess for inviting me to give a presentation at the workshop on The Philosophy of Virtuality and for inviting me to develop the presentation into this article. Their encouragement and helpful suggestions have been invaluable.

References

Beck, U. (1992). *Risk Society: Towards a New Modernity*. London: Sage Publications.

Bok, S. (1989). *Lying: Moral Choice in Public and Private Life*. New York: Vintage Books.

Briggle, A. (2008). Real Friends: How the Internet can Foster Friendship. *Ethics and Information Technology*, *10*, 71–79.

Caygill, H. (1995). *A Kant Dictionary*. Oxford: Blackwell Publishers.

Cocking, D. & Matthews, S. (2000). Unreal Friends. *Ethics and Information Technology*, *2*, 223–231.

Dibble, J. (1993). A Rape in Cyberspace. *The Village Voice*, December 23, 1993.

Elster, J. (2007). *Explaining Social Behaviour*. Cambridge: Cambridge University Press.

Grimen, H. (2009). *Hva er tillit* [What is trust]? Oslo: Universitetsforlaget.

Hardin, R. (1996). Trustworthiness. *Ethics*, 107, 26–42.

Hardin, R. (2006). *Trust*. Cambridge: Polity Press.

Herman, B. (1996). *The Practice of Moral Judgment*. Cambridge, MA.: Harvard University Press.

Kant, I. (1965). *Grundlegung zur Metaphysik der Sitten* [*Groundwork of the Metaphysics of Morals*]. Hamburg: Felix Meiner Verlag.

Kant, I. (1968). *Schriften zur Ethik und Religionsphilosophie* [*Writings on Ethics and Philosophy of Religion*] *2*. Frankfurt am Main: Suhrkamp Verlag.

Kant, I. (1983). *Anthropologie in pragmatischer Hinsicht* [*Anthropology from a Pragmatic Point of View*]. Stuttgart: Reclam Verlag.

Kant, I. (1987). *Critique of Judgment*. Indianapolis: Hackett Publishing Co.

Kant, I. (1996). *The Metaphysics of Morals*. Cambridge: Cambridge University Press.

Kant, I. (1997). *Lectures in Ethics*. Cambridge: Cambridge University Press.

Kierkegaard, S. (1994). *Afsluttende Uvidenskabelig Efterskrift* [*Concluding Unscientific Postscript*]. København: Gyldendal.

Korsgaard, C. (1996). *Creating the Kingdom of Ends*. Cambridge: Cambridge University Press.

Louden, R. B. (2000). *Kant's Impure Ethics: From Rational Beings to Human Beings*. Oxford: Oxford University Press.

Luhmann, N. (1989). *Vertrauen: Ein Mechanismus der Reduktion sozialer Komplexität* [*Trust: A mechanism for reducing social complexity*]. Stuttgart: Ferdinand Enke Verlag.

Myskja, B. (2008). The Categorical Imperative and the Ethics of Trust. *Ethics and Information Technology*, *10*, 213–220.

Neiman, S. (2002). *Evil in Modern Thought: An Alternative History of Philosophy*. Princeton, NJ: Princeton University Press.

Nissenbaum, H. (2001). Securing Trust Online: Wisdom or Oxymoron? *Boston University Law Review*, *81*, 101—131.

Nussbaum, M. C. (1990). *Love's Knowledge: Essays on Philosophy and Literature*. Oxford: Oxford University Press.

O'Neill, O. (2002). *Autonomy and Trust in Bioethics*. Cambridge: Cambridge University Press.

Turkle, S. (1995). *Life on the Screen: Identity in the Age of the Internet*. New York: Simon & Schuster.

Weckert, J. (2005). Trust in Cyberspace. In R. J. Cavalier (Ed.), *The Impact of the Internet on Our Moral Lives*, 95–117. Albany: State University of New York Press.

Wittgenstein, L. (1992). *Philosophical Investigations*. Oxford: Blackwell Publishers.

SECTION III

Applications/Implications

CHAPTER EIGHT

Virtual Child Pornography

Why Images Do Harm from a Moral Perspective

Litska Strikwerda

Introduction

There seems to be an international trend that seeks to ban child pornography entirely. This is evidenced by the recent adoption of the Optional Protocol to the UN Convention on the rights of the child, on the sale of children, child prostitution and child pornography (adopted and opened for signature, ratification and accession by General Assembly resolution A/RES/54/263 of 25 May 2000), the recent European Commission initiative on combating sexual exploitation of children and child pornography (Council Framework Decision 2004/68/JHA of 22 December 2003), and the adoption of the Convention on Cybercrime (Council of Europe, ETS No. 185, Budapest, 23 November 2001). All tend to criminalize the electronic possession, distribution and production of child pornography.

John Stuart Mill's liberalism provides for the 'harm principle' as the 'moral basis' for criminal law (Bedau 1974). According to Mill (as cited in Dworkin, 1972) the harm principle entails:

> That the only purpose for which power can be rightfully exercised over any member of a civilised community, against his will, is to prevent harm to others. (Mill, 1859, 135)

With reference to child pornography there seems to be a discussion going on whether it can constitute harm, since it consists of 'just images'. Therefore, the questions arise whether the harm principle applies to the prohibition of child pornography and, if not, what is the moral basis for prohibition. My aim is to answer these questions.

I will begin with a (legal) definition of child pornography. Discussing different categories of child pornography, I will highlight the question as to whether virtual child pornography (child pornography consisting entirely of computer-generated images) falls under the harm principle. I will suggest then that virtual child pornography is a 'victimless crime' (as described by Bedau) to which the harm principle does not apply. According to Bedau, the criminalization of victimless crimes is based on either paternalism or moralism. Therefore I will continue by exploring whether the prohibition of virtual child pornography is based on paternalism or moralism. Ultimately drawing from the positions of virtue ethics and feminism, I will argue that these provide a basis through moralism to justify the criminalization of virtual child pornography in liberal societies.

Child pornography: Definitions

The question what constitutes child pornography is a complex one. Legal definitions of both 'child' and 'pornography' differ globally (Quayle & Taylor 2002, 865). As I see it, there are three categories of child pornography.

1. **Photographs** showing a minor, or a person appearing to be a minor, engaged in sexually explicit conduct.
2. **'Pseudo photographs'**: photographs of actual children which have been manipulated or 'morphed' to make it appear they are engaged in sexual activity.
3. **Entirely computer-generated images** that do not depict actual children at all.

All categories seem to be covered by the following definition of 'child pornography', as provided for by article 9 (2) of the Convention on Cybercrime (CETS No. 185)[1]:

> ...the term 'child pornography' shall include pornographic material that visually depicts:
> a a minor engaged in sexually explicit conduct;
> b a person appearing to be a minor engaged in sexually explicit conduct;

c realistic images representing a minor engaged in sexually explicit conduct.

The third paragraph of the above-mentioned article defines the term 'minor' as '*all persons under 18 years of age.*' And according to the Council of Europe Convention on Cybercrime Explanatory Report (CETS no. 185) 'sexually explicit conduct' covers:

> a) sexual intercourse, including genital-genital, oral-genital, anal-genital or oral-anal, between minors, or between an adult and a minor, of the same or opposite sex; b) bestiality; c) masturbation; d) sadistic or masochistic abuse in a sexual context; or e) lascivious exhibition of the genitals or the pubic area of a minor. It is not relevant whether the conduct depicted is real or simulated. (§100)

The Convention on Cybercrime has been ratified by all members of the Council of Europe as well as the United States, Canada, Japan and some other non-member countries. This means the above-mentioned definition of 'child pornography' is widespread. Therefore, I will stick to this definition for the purpose of this paper.

Criminalization of child pornography

Various aspects of the electronic possession, distribution and production of child pornography as defined in article 9 (2) of the Convention on Cybercrime have been criminalized by article 9 (1) of the Convention on Cybercrime. Most of the signatory states to the Convention had already established this conduct as criminal offences under their domestic law before, but after ratification they shaped their definition of child pornography after the aforementioned definition (see for example the Dutch Criminal Code provision on child pornography: article 240b Wetboek van Strafrecht). In sum, the possession, distribution and production of child pornography as defined in article 9 (2) of the Convention on Cybercrime are seen as criminal offences worldwide.

According to the Explanatory Report (§91) article 9 of the Convention on Cybercrime '*seeks to strengthen protective measures for children, including their protection against sexual exploitation [...]*' by criminalizing child pornography. This aim has been explained as follows. The criminalization of the first category of child pornography '*focuses more directly on the protection against child abuse*' (Expl. Report, §102). The criminalization of the second and third category aims at

> providing protection against behaviour that, while not necessarily creating harm to the 'child' depicted in the material, as there might not be a real child, might be used to encourage or seduce children into participating in such acts, and hence form part of a subculture favouring child abuse. (Expl. Report, §102)

These arguments both seem to appeal to the 'harm principle', the first-mentioned more directly than the second. In conclusion, the assumption made in the Convention on Cybercrime is that the criminalization of the possession, distribution and production of child pornography is legitimate, because this conduct harms children.

Do all categories of child pornography harm children?

The question arises whether the above-mentioned assumption is correct. With regard to the first category of child pornography at least, there seems to be consensus that this type of child pornography indeed harms children. It is important to note here that in sex contacts between adults and children, mutual consent is in general assumed to be absent (Moerings, 1999, 190). The harm done to children by the first category of child pornography can therefore be argued as follows.

Child pornography is 'a recording of sexual abuse':

> In the production of this material, pornography and sexual abuse overlap, and any conceivable legitimation has been eradicated. (Boutellier, 2000, 455)

Following this argument, the possession and distribution of this type of child pornography can also be considered harmful to children. After all, the 'consumption' of child pornography of this sort supports the 'market' for it and therefore indirectly causes the sexual abuse of children that comes with the production (Moerings, 1999, 191).

The criminalization of the second category of child pornography seems to be generally accepted because of 'the possibility that harm has been done.' Due to the worldwide range of the Internet, it is impracticable for authorities to track down the child abused for every single child pornographic image found, thereby proving 'real' harm has been done. And due to the advanced computer techniques used, it is sometimes even impossible to say whether there has been an actual child involved in the production or just a photograph of an actual child. The possibility an actual child has been abused for the production of this type of child pornography and real harm has been done, can therefore never be completely ruled out.

In conclusion, the nature of this argument is a more pragmatic one. It makes it easier for authorities to combat online child pornography effectively, because they do not need to prove an actual child has been involved in the production in order to convict a person for the possession or distribution of it (Kamerstukken II 2000/2001, 27 745/3, 4).

> There is also another argument to be thought of here. If one morphs an 'innocent' picture of an actual child into child pornography, the child depicted can be harmed, although not sexually abused. The harm done is then contained in the violation constituted to the privacy of this child (Levy, 2002, 319).

A recent judgment of the European Court of Human Rights (ECtHR) is of importance here. The facts of the case were the following. An unknown person placed an advertisement on an online dating site in the name of a 12-year-old boy, without his knowledge. The advertisement mentioned his age, gave a detailed description of his physical characteristics, a link to his web page, as well as his telephone number, which was accurate save for one digit. In the advertisement, it was claimed that he was looking for an intimate relationship with a boy of his age or older '*to show him the way.*' The boy became aware of the announcement when he received an e-mail from a man, offering to meet him and '*then to see what you want.*'

The ECtHR ruled the boy's right to privacy (article 8 of the European Convention on Human Rights and Fundamental Freedoms) had been violated:

> There is no dispute as to the applicability of Article 8: the facts underlying the application concern a matter of 'private life', a concept which covers the physical and moral integrity of the person. [...] The Court would prefer to highlight these particular aspects of the notion of private life, having regard to the potential threat to the applicant's physical and mental welfare brought about by the impugned situation and to his vulnerability in view of his young age. (ECtHR, December 2, 2008, *K.U. v. Finland*, Appl. No. 2872/02, para. 41)[2]

With regard to the third category of child pornography, however, there seems to be a discussion going on about the question whether it harms children. Most important here is the legal origin of article 9 of the Convention on Cybercrime itself.

The definition of child pornography used in the article mentioned above largely resembles the definition as was provided for by Sec. 2256 of the US Child Pornography Prevention Act. This act has been replaced by another act (Act on Sexual Exploitation and other Abuse of Children) including a different definition (Sec. 2252A Act on Sexual Exploitation and other Abuse of Children), because the Child

Pornography Prevention Act was found unconstitutional by the US Supreme Court, partly on the grounds that the third category of child pornography distinguished does not harm actual children.

The Supreme Court stated:

> Virtual child pornography is not 'intrinsically related' to the sexual abuse of children. While the Government asserts that the images can lead to actual instances of child abuse, the causal link is contingent and indirect. The harm does not necessarily follow from [it], but depends upon some unquantified potential for subsequent criminal acts. (*Ashcroft v. Free Speech Coalition*, April 16, 2002, No. 00-795)

In conclusion, the argumentation of the Supreme Court here is twofold. To begin with, the Supreme Court confirms the first assumption made in the Explanatory Report to the Convention on Cybercrime as quoted earlier: that the production and related distribution and possession of entirely computer-generated ('virtual') child pornography does not constitute any victims of sexual abuse and therefore does not harm any children *directly*. Secondly, the Supreme Court doubts the correctness of the second assumption made in the Explanatory Report: that virtual child pornographic images could lead to actual instances of child abuse. The Supreme Court therefore holds the opinion virtual child pornography does not harm children *indirectly* either.

Virtual child pornography as a 'victimless crime'

The first argument of the Supreme Court, which confirms the assumption made in the Explanatory Report to the Convention on Cybercrime, seems correct to me and can be analysed as follows. The production, distribution and possession of virtual child pornography is seen as a 'victimless' crime (Moerings, 1999, 192).

Bedau has described the concept of a 'victimless crime' as follows:

> An activity is a victimless crime if and only if it is prohibited by the criminal code and made subject to penalty or punishment, and involves the exchange or transaction of goods and services among consenting adults who regard themselves as unharmed by the activity and, accordingly, do not willingly inform the authorities of their participation in it. (1974, 73)

He thinks for instance of prohibitions on prostitution and gambling as victimless crimes (Bedau, 1974, 61, 85).

The prohibition on the production, distribution and possession of virtual child pornography seems to meet the requirements set by the above-mentioned definition as well. First of all, the distribution of virtual child pornographic images can be seen as an 'exchange of goods'. The distribution leads to possession. And the production of virtual child pornography is also inevitably related to this exchange. On the one hand, no distribution and possession would be possible without production; on the other hand, distribution and (the demand for) possession create the market for production.

It is the production of virtual child pornography that makes the difference with the two other categories of child pornography. Since they are entirely computer-generated, the virtual child pornographic images exchanged can be seen as neutral 'goods', instead of a (possible) recording of sexual abuse of children. In the production of virtual child pornography, there cannot be 'children of flesh and blood' involved whose rights can be violated, as opposed to the two other forms of child pornography. Therefore, the production of virtual child pornography does not seem to harm any children directly, nor can the possession and distribution of it directly be linked to any harm to children (Bedau, 1974, 63).

Secondly, if I limit myself to adults who intentionally exchange virtual child pornographic images, they can be considered to meet the requirement of consent as well. Finally, the adults consenting to the exchange of virtual child pornographic images do not seem to regard themselves as harmed by this activity. Many offenders would argue the opposite and state that looking at child pornographic images provides a safe outlet for feelings that otherwise would lead to a 'contact offence' (O'Brien & Webster, 2007, 238).

A victimless crime does not need to be harmless. Bedau states:

> The presence of the [...] defining attributes of victimless crimes simply cannot guarantee, either conceptually or empirically, that every such crime is harmless to the participants. (1974, 74)

After all, the question whether harm is done is made subject to the self-judgment of the participants. And their judgment might be objectively seen as incorrect (Bedau, 1974, 76).

Bedau also points out that the concept of a victimless crime as described above includes

> the tacit assumption that if the participants in the activity consent to it and judge themselves unharmed by engaging in it, nobody else can be injured by it, either; and if that is so, society has no right to interfere by prohibiting the activity and subjecting it to penal sanctions. (1974, 75)

And this assumption is according to Bedau also incorrect. A victimless crime might affect other people than just the participants in it, therefore society can have the right to interfere by prohibiting the activity (Bedau 1974, 75).

Paternalism as a moral basis for the criminalization of victimless crimes

According to Bedau (1974), the criminalization of victimless crimes is based on either paternalism or moralism. Paternalism can be defined as follows:

> ...the interference with a person's liberty of action justified by reasons referring exclusively to the welfare, good, happiness, needs, interests or values of the person being coerced. (Dworkin, 1972, 65)

As opposed to the harm principle, which is about harm to *others*, paternalism is about harm to the *self*.

Paternalism is based on the assumption that society also needs to be protected from people who willingly take risks for themselves and others (Bedau 1974, 71). Mill (as cited in Dworkin, 1972) explicitly singled out paternalism as a moral basis for criminal law. He stated:

> His own good, either physical or moral, is not a sufficient warrant. (Mill 1859, 135)
>
> According to Feinberg however, 'presumptively nonblamable paternalism' entails:
>
> ...defending relatively helpless or vulnerable people from external dangers, including harm from *other* people when the protected parties have not voluntary consented to the risk, and doing this in a manner analogous in its motivation and vigilance to that in which parents protect their children. (1986, 5)

This can also be called the *parens patriae* principle, which is vital to the doctrine of (criminal) law (Feinberg, 1986, 6).

The *parens patriae* principle could be a solid moral basis for the criminalization of the production, distribution and possession of virtual child pornography if it could be proven that virtual child pornographic images could encourage or seduce children into participating in sexual contacts with adults. The moral basis for prohibition would then be situated in the protection of children against the aforementioned seduction or encouragement, thereby defending them from harm from paedophiles.

This would fit perfectly into the framework of traditional protections which the criminal law has provided for children '*against exploitation, manipulation and injury at the hands of adults*' (Bedau, 1974, 86, emphasis added, LS). Statutory rape, for instance, traditionally finds its legitimation in these protections (Moerings 1999, 190).

Paternalism, in the sense in which it is a proposed principle for the moral legitimization of criminal legislation with regard to adults who are not relatively helpless or vulnerable, Feinberg (1986, 6f.) calls 'legal paternalism'. Legal paternalism consists of

> criminal prohibitions that can be defended, at least initially, on two distinct grounds, both the need to protect individuals from the harmful consequences of their own acts and the need to prevent social harm generally. (Feinberg, 1986, 21f.)

Legal paternalism could be the moral basis for the criminalization of virtual child pornography if proven that it encourages or seduces paedophiles to commit child abuse.

In conclusion, if the doubts of the US Supreme court as to whether virtual child pornography could lead to actual instances of child abuse (a doubt that, as noted above, rejects the assumption at work in the Explanatory Report to the Convention on Cybercrime) turn out to lack justification—then both forms of paternalism can provide a moral basis for criminalization of virtual child pornography. Examining this foundational assumption will be the focus of the next two sections.

Does virtual child pornography encourage or seduce children into participating in sexual contacts with adults?

Investigators have found links between young people who watch pornographic images and their attitudes toward sex (DeAngelis 2007; Rutgers Nisso Groep/ Nederlands Jeugdinstituut/ Movisie, 2008; Movisie 2009). It has been suggested that the younger the child is, the more influence these images have (Movisie 2009, 70).

It does not seem likely that children would deliberately search the Internet themselves for virtual child pornographic images. But child pornographic images might well be used by offenders to 'groom' children to take part in sexual acts (Johnson & Rogers, 2009, 77). Showing them to a child could be used to encourage participation, stimulate arousal or as an example of what the offender wishes the child to do (Quayle & Taylor 2002, 866). And the effect of the images on children could be that they come to think the activity must be acceptable, since others have engaged in it (Levy, 2002, 320).

Research suggests many perpetrators arrested for Internet sex crimes had child pornography in their possession (Johnson & Rogers, 2009, 77). And it has been found that quite a number of arrested child pornography downloaders admit to having abused an actual child as well (University of Groningen/Utrecht University, 2004–2005).

There is no evidence however, that paedophiles do frequently make use of virtual child pornographic images in the way as described above. And more importantly, they can and do make use of a lot other means to groom children to take part in sexual acts, such as drugs, alcohol, toys, money or force (Levy 2002, 320).

As a Dutch court has pointed out, it makes a difference if the virtual pornographic images are specifically aimed at children. A Dutch national was recently convicted for the possession of an entirely computer-generated film of this nature. The film was titled *Sex Lessons for Young Girls/Lessons Jerking and Facial* and showed a virtual girl about 8 years of age engaged in sexually explicit conduct with a man. The girl depicted is smiling, the man applauds for her and colorful balloons appear.

The court argued for the verdict as follows. The persons appearing in the film do not seem real to adults, but do so to the average child. Due to the 'instructional' nature of the film and the colorful framing, it seems to be aimed at children. Therefore, it could be used to encourage or seduce children into participating in such acts (Rb.'s-Hertogenbosch, February 4, 2008, LJN: BC3225).

The verdict can be further explained as follows. Since the child pornographic images at stake were specifically aimed at children, the intent to use them to groom children into participating in sexual acts with adults followed directly from them.

In this case, but only in this case, the *parens patriae* principle provides for a solid ground for prohibition. The production, distribution and possession of virtual child pornographic images of this nature need to be prohibited in order to protect children against exposure to them and the risk of seduction or encouragement to participate in harmful sexual contacts with adults that comes with that.

With reference to all other kinds of virtual child pornographic images, however, the causal link between them and actual instances of abuse is, as the US Supreme Court stated, '*contingent and indirect.*' Therefore, this form of paternalism cannot provide for a solid ground for criminalization in general.

Does virtual child pornography encourage or seduce paedophiles to commit child abuse?

As stated earlier many offenders argue that virtual child pornography has a positive rather than a negative effect on them, because looking at such images provides a safe outlet for feelings that otherwise would lead to a 'contact offence.' However, there is little evidence on this. It can even be argued that the reverse is true (O'Brien & Webster, 2007, 238).

Some psychologists are of the opinion that watching child pornographic images encourages paedophiles to abuse a child. The Dutch psychologist Buschman for instance found in a study of 23 arrested child pornography downloaders that 6 of them admitted under polygraph to have abused an actual child (University of Groningen/Utrecht University 2004–2005). Other research seems to support these findings (see for an overview: O'Brien & Webster, 2007, 238–239).

However, the research performed is criticized for using a small sample. There seems to be too little evidence yet to prove that a direct causal link exists between virtual child pornographic images and child abuse in this way (Levy, 2002, 320). There might be an indirect link though.

Prior to the Internet, paedophiles remained a relatively isolated group, but this new technology has enabled them to form social networks online, which are called virtual communities or subcultures. The Internet provides them with a medium of exchange, not only of child pornographic images, but also of ideas. Due to its anonymity, the Internet provides for a relatively safe environment for this exchange of this illegal or at least generally disapproved content (Quayle & Taylor, 2002, 867).

According to the Explanatory Report to the Convention on Cybercrime the subculture thus formed favours child abuse. This has been explained as follows:

> It is widely believed that such material and on-line practices, such as the exchange of ideas, fantasies and advice among paedophiles, play a role in supporting, encouraging or facilitating sexual offences against children. (§ 93)

This consideration seems to appeal to the process of 'group polarization' as described by Sunstein. According to Sunstein, the Internet gives people the opportunity to meet 'similar identities' (like-minded people), in a way that was never possible before:

> The internet gives you the opportunity to meet other people who are interested in the same things you are, no matter how specialized, how weird, no matter how big or how small. (2001, 54)

Sunstein further observes that like-minded people are more likely to convince each other with their arguments online:

> If identity is shared, persuasive arguments are likely to be still more persuasive; the identity of those who are making them gives them a kind of credential or boost. (2001, 70f.)

This can lead to the process of 'group polarization':

> ...after deliberation, people are likely to move toward a more extreme point in the direction to which the group's members were originally inclined. With respect to the internet [...] the implication is that groups of like-minded people, engaged in discussion with one another, will end up thinking the same thing that they thought before—but in a more extreme form. (Sunstein, 2001, 65)

Sunstein claims there are two main explanations for the process of group polarization. The first emphasizes the role of above-mentioned *persuasive arguments*. On this account, the central factor behind group polarization is the existence of a *limited argument pool*, which is skewed in a particular direction. The tendency of online discussion groups is to entrench and reinforce preexisting positions, often resulting in extremism (Sunstein, 2001, 67–68).

The second explanation appeals to the idea of *the spiral of silence*. In groups people want to be perceived favorably by other group members and also want to perceive themselves favorably. Once they hear what others think, they often adjust to the most dominant position in the group. Critical minorities silence themselves (Sunstein, 2001, 68–69).

According to Quayle & Taylor (2002) and O'Brien and Webster (2007) there can indeed be recognized a process of group polarization in the virtual communities formed by paedophiles. The following four characteristics of cognitive distortions are common in those who download child pornography.

First of all, the *justification* of the behavior, because child pornographic images are 'just images' and do not involve contact with an actual child. Secondly, the *normalization* of the behavior: child pornographic images could validate and justify the behaviour of those with a sexual interest in children, providing proof to them that the existence of such material shows that their behaviour is not abnormal but is shared by thousands of others. Also, most images portray compliant, often smiling children, which could further contribute to the sense of appropriateness and validation. Thirdly, the *objectification* of the images: through a process of collecting, the downloader distances himself from the illegal content. The images are used as 'a medium of exchange' in which the images in themselves act as a form of currency, thereby legitimizing activity and creating social cohesion. And finally, the *justification of other forms of engagement* with the images or, on occasion, real children through colluding in a social network (Quayle & Taylor, 2002, 866–869; O'Brien & Webster, 2007, 241).

In conclusion, paedophiles engaged in the exchange of child pornographic images will end up thinking of children in the same way as they thought of them before: as sex objects, but in a more extreme form: that it is justified, normal and not harmful to think of children this way. And they might even come to think that actual child abuse is justified.

However, the process of group polarization as described above does not only consist of the exchange of images. This exchange is likely to be accompanied by an exchange of speech (Levy, 2002, 321). Without the exchange of speech, the fourth cognitive distortion distinguished does not seem possible at all.

And even if the exchange of virtual child pornographic images solely would lead to a process of group polarization as described above, I doubt whether that would be enough ground for criminalization. After all, it would provide for an indirect link to an indirect harm. The above-mentioned definition of legal paternalism however, seems to require a direct link to harm, either to the self or society at large. And as the Supreme Court stated, the causal link between virtual child pornographic images and actual instances of abuse cannot be '*contingent and indirect*' if the prohibition on the production, distribution and possession of such images is to be legitimated on the ground that this conduct harms children.

In conclusion, legal paternalism could provide for a solid moral basis for the prohibition on the production, distribution and possession of virtual child pornography if more research would provide for more evidence in the future. The moral ground for prohibition would then be based in the protection of paedophiles individually and society at large from the harmful effects that exposure to virtual child pornographic images has on them (it makes them commit the crime of sexual abuse of children). However, research has not come that far yet.

'Thinking outside the box': The virtue ethics point of view

In sum, due to a lack of reliable data it cannot be proven virtual child pornographic images encourage or seduce children into participating in sexual contacts with adults, nor that they encourage or seduce paedophiles to commit child abuse. Therefore, paternalism of any kind, as an alternative for the harm principle, cannot provide for a moral basis for the prohibition on the production, distribution and possession of virtual child pornography either.

As opposed to liberalism and paternalism, the virtue ethics approach focuses not on acts themselves, but on the *motives* they are performed with. As will become clear later on, virtue ethics provides for an alternative for liberalism and paternalism in this discussion, because from this perspective it is not necessary to prove a causal link between virtual child pornographic images and actual instances of abuse to consider them as harmful.

Alasdair MacIntyre, who specifically relates virtues to the social or professional roles people have in society, provides the following definition of a 'virtue':

> A virtue is an acquired human quality the possession and exercise of which tends to enable us to achieve those goods which are internal to practices and the lack of which effectively prevents us from achieving any such goods. (1984, 191)

By a 'practice' MacIntyre means:

> any coherent and complex form of socially established cooperative human activity through which goods internal to that form of activity are realized in the course of trying to achieve those standards of excellence which are appropriate to, and partially definitive of, that form of activity with the result that human powers to achieve excellence, and human conceptions to the ends and goods involved, are systematically extended. (1984, 187)

Internal goods, at last, he describes as those goods that can only be achieved by engaging in the practice (MacIntyre 1984, 188)

In her virtue ethical theory of "Better Sex" (1975) Sara Ruddick has analyzed the practice of sex. According to Ruddick the best sex is 'complete sex'. Whether a sexual act is complete depends upon the relation of the participants to their own and each other's desire. The precondition of complete sex is the embodiment of the participants. Through complete sex acts the internal good of '*reflexive mutual recognition of desire by desire*' can be achieved (Ruddick 1975, 87–89). Reflexive mutual recognition of desire by desire entails the following:

> ...in complete sex two persons embodied by sexual desire actively desire and respond to each other's active desires. [...] complete sex is reciprocal sex. The partners [...] are equal in activity and responsiveness of desire. (Ruddick 1975, 90)

According to Ruddick (1975, 86) a sex act is morally superior if it is more virtuous than another sex act or more likely to lead to a virtue. Complete (embodied) sex acts are morally superior to other sex acts because

> they are conductive to emotions that, if they become stable and dominant, are in turn conductive to the virtue of loving; and they involve a pre-eminently moral virtue-respect for persons. (Ruddick 1975, 98)

This does not mean incomplete sex acts do necessarily involve moral disrespect for persons, however (Ruddick 1975, 100). Ruddick states:

> ...complete sex acts are superior to incomplete ones [...] because they involve a kind of 'respect for persons' in acts that are otherwise prone to violation of respect for, and often violence to, persons. [...] Any sexual act that is pleasurable is *prima facie* good, though the more incomplete it is—the more private, essentially autoerotic, unresponsive, unembodied, passive, or imposed—the more likely it is to be harmful to someone. (1975, 101)

In conclusion, Ruddick thinks the virtue of respect for persons enables the participants to achieve the good of reflexive mutual recognition of desire by desire (reciprocity) which is internal to the practice of complete sex. The precondition of complete sex is the embodiment of the participants. Sex practices through which no reciprocity is realized, do not necessarily involve disrepect for persons, but are likely to be harmful to the virtue of respect for persons or actual persons.

The feminist critique of pornography

Feminists in general endorse strong government policies to fight what they see as demeaning cultural images of women and children (Kymlicka, 2001, 393). In their fight, they use arguments that resemble and extend Ruddick's theory. MacKinnon, for instance, argues against pornography as follows:

> Pornography sexualises rape, battery, sexual harassment, prostitution and child sexual abuse; it thereby celebrates, promotes, authorizes and legitimises them. More generally, it eroticizes the dominance and submission that is the dynamic common to them all. (1992, 461)

First of all, MacKinnon describes the sex acts depicted by pornographic images as 'rape, battery, sexual harassment, prostitution or child sexual abuse.' Ruddick would say that these are all examples of incomplete—unresponsive, passive and imposed—sex acts. She condemns these sorts of acts, insofar as they reduce others to sexual objects that are treated (in Kantian terms) solely as a means to fulfill one's own ends, because thereby the primary virtue of respect for persons is directly violated.

Secondly, MacKinnon claims pornographic images *eroticize* the dominance and submission that is common to the sex acts they depict. Ruddick would describe this 'dominance and submission' as a lack of the reciprocal recognition of the other as a person (in Kantian terms an end rather than a means to one's own sexual ends) that characterizes complete sex and its fostering of the primary virtue of respect for persons.

The concept of *eroticization* can be explained as follows. 'Mainstream' pornographic images in general represent male dominance and female subordination (there is no reciprocity between them). They thereby affirm the place women have in the social structure, '*not merely expressing subordination, but in part constituting it*' (Itzin, 1992, 61, emphasis added, LS). In sum:

> Women are […] subordinated in pornography, and women are subordinated as a result of the use of pornography. (Itzin, 1992, 67)

Here, first of all, a clear distinction is made between the sex act *depicted* by pornographic images and the sex act of *watching* pornographic images. As stated above, Ruddick would consider the sex acts depicted by pornographic images as incomplete and condemn them. Applying Ruddick's theory to the sex act of watching pornography, it cannot be complete either, since it lacks embodiment. Therefore, no reciprocity can occur. This leads to the assumption that the sex act of watching pornography is likely to be harmful to the virtue of respect for persons or actual persons. The essentially autoerotic, unresponsive and passive nature of this sex act adds to this assumption. In sum, one could say watching pornography is 'the incomplete sex act of watching incomplete sex acts.'

As Itzin explains, this incomplete sex act of watching incomplete sex acts is not only harmful to the virtue of respect for persons, but also to actual women. Pornographic images present to men '*how it is permissible to look at and to see women*' (Itzin, 1992, 67, emphasis added, LS). They learn '*to see women in terms of their sexuality and sexual inequality*' (Itzin, 1992, 67, emphasis added, LS).). As a result of watching pornography:

> …the ideas of pornography enter the imagination and are transformed into attitudes and behaviour: into actions. (Itzin, 1992, 68)

Itzin seems to claim here that watching pornography infringes with the sexual mentality based on equality (Boutellier, 2000, 448). This sexual mentality is rooted in a broader 'equality norm', which has been formulated in amongst others the UN Universal Declaration of Human Rights (United Nations, 1948). Article 1 of the Declaration reads as follows:

> All human beings are born free and equal in dignity and rights.

In conclusion, pornographic images depict incomplete, non-reciprocal sex acts, in which women are unequal to men. They thereby violate the sexual mentality based on the equality norm. The sex act of watching such images can lead to *actual* violations of the sexual mentality based on the equality norm, because the message sent by pornography ('non-reciprocal sex acts are erotic!') influences attitudes and behaviour towards *actual women*.

The linkage between sexual attitudes towards women and consuming pornography that eroticizes non-reciprocal sex has been confirmed by research. It has been demonstrated there are connections between 'an excess of one-sided, gender-stereotyped, sexual content' in the media on the one hand and the sexual and psychosocial well-being of young people on the other. It seems to influence their sexual experience and the likelihood of their accepting and/or engaging in sexual harassment (Rutgers Nisso Groep / Nederlands Jeugdinstituut / Movisie 2008).

So it is that Itzin argues for the 'elimination' of pornography. She concludes:

> The elimination of pornography is an essential part of the creation of genuine equality for women—and for men. (Itzin, 1992, 70)

MacKinnon is more modest. She states:

> Women will never have that dignity, security, compensation that is the promise of equality *so long as the pornography exists as it does now*. (MacKinnon, 1992, 486: emphasis added, LS)

The latter suggests that pornography does not need to disappear in order to protect a sexual mentality based on the equality norm. Instead, it would also be possible to change the sex acts depicted to more complete, reciprocal and thus equal ones.

(Virtual) child pornography and equality

The possibility of changing 'mainstream' pornography in the direction of greater equality, reciprocity, and thus completeness, however, highlights a fundamental difference between 'mainstream' pornography and child pornography:

> ...child pornography, actual or virtual, cannot depict children as equal participants in sexual activity with adults [...]. Children are not equal [...]. For that reason, sexualising children for adult viewers is necessarily sexualising inequality. Child pornography is an extension of mainstream sexual relations, which are contingently unequal, into new arenas. (Levy, 2002, 322)

Therefore, child pornography can be seen as '*the culmination of a sexualised culture*':

> Child pornography is the most extreme consequence of the schizophrenic mindset of pornographic permissiveness and the equality norm. This form of pornography shows utter contempt for our cultural idea of equality. (Boutellier, 2000, 455–456)

With reference to *virtual* child pornography there is another reason why it is impossible to change the sex acts depicted to more complete, reciprocal and thus equal ones. Here, another comparison with mainstream pornography needs to be made.

Mainstream pornographic images do not only influence the way women are viewed by men, but also the way women view themselves. According to research self-objectification of girls is mainly related to them (Rutgers Nisso Groep / Nederlands Jeugdinstituut / Movisie 2008). Itzin (1992, 62) refers to pornographic images as '*mirror images*' in this context.

However, in the pornographic images that are manifest in the media, a highly overrated beauty ideal is presented (Rutgers Nisso Groep / Nederlands Jeugdinstituut / Movisie 2008). Most of them are photo-shopped in such a way that no 'natural woman' could reach the standards set. A much-discussed Dutch documentary (Bergman, 2007) shows US women, including a 14-year-old girl, who undergo plastic surgery to look like the women they have seen depicted in such images. This is not because they want to look like porn stars, but because they consider those images as realistic and think that they themselves are abnormal. The cosmetic industry and the porn industry seem to have become intertwined.

Applying Ruddick's theory, one could say that the sex acts *depicted* by pornographic images as described above lack embodiment. They do not show the participants engaged in a sex act, but a photo-shopped version of them. Since embodiment is the precondition of complete sex, such images do not depict a complete, reciprocal sex act.

In comparison, I would say that virtual child pornographic images have even further drifted apart from the 'embodied reality'. They are not just photo-shopped: there has not even been an actual child involved in the production at all. Therefore, virtual child pornographic images lack the precondition of complete sex as well.

In the production of mainstream pornographic images as described above, the 'photo-shopping' could easily be left out and the effect would be that these images show natural features again. However, the unembodied nature of virtual child pornographic images cannot be changed, since they are *entirely* computer-generated.

In conclusion, child pornographic images depict incomplete, non-reciprocal sex acts. They thereby violate a sexual mentality based on the equality norm. As opposed to the sex acts depicted by mainstream pornographic images, the incomplete nature of the sex acts depicted by child pornographic images cannot be changed, due to the fact that children *are* unequal to adults and could therefore never be engaged in complete, reciprocal sex acts with them. With reference to virtual child pornographic images a specific reason needs to be added: they necessarily lack embodiment, which is the precondition of complete sex. Therefore, (virtual) child pornographic images necessarily eroticize inequality.

So, while research fails to conclusively demonstrate a linkage between consumption of child pornography and attitudes and behavior towards *actual* children—there does seem to be good research supporting the virtue ethics view that the consumption of pornography that eroticizes inequality influences attitudes and behavior towards *actual* women. If the assumption that (virtual) child pornography necessarily eroticizes inequality is correct, it may be that the wish to achieve equality in our sexual attitudes and ethics is (part of) the drive behind the international trend to ban the production, distribution and possession of child pornography (Boutellier, 2000, 456.).

Moralism as the moral basis for the prohibition on the production, distribution and possession of virtual child pornography

In reply to feminists, liberals would typically say that while pornography may offer a false representation of (women's) sexuality, that is not a sufficient ground for legally restricting it (Kymlicka, 2001, 393). MacIntyre explains:

> ...it is on the liberal view no part of the legitimate function of government to inculcate any one moral outlook. (1984, 195)

If the law does inculcate moral outlook, liberals call this 'moralism'. Bedau provides for the following definition of moralism:

> ...the policy of using the criminal law to curb a person's freedom of action not on the ground that it is unfairly harmful to others (for this is already prohibited by liberalism), nor on the ground that it is an irrational harm to the participant (this is already prohibited by paternalism), but solely on the ground that it is offensive, degrading, vicious, sinful, corrupt or otherwise immoral. (1974, 89–90)

The above-mentioned point of view needs some clarification though. After all, the entire criminal code seems to express the community's ideas of morality, or at least of the most egregious immoralities. As Bedau explains, however, most crimes are the kind of *harmful* immoralities which Mill's 'harm principle' was designed to prohibit. Others can be legitimized by paternalism. Moralism refers to the criminalization of immoralities that are not harmful to others, nor to the self (Bedau 1974, 91–92).

Bedau explains that the aforementioned immoralities

> can cause offense to persons [...] by virtue of flouting deeply held convictions, and that in being offensive these activities are harmful or something sufficiently akin to harmfulness as to be virtually indistinguishable from it. (1974, 97)

Virtual child pornography seems to be such an immorality. As the virtue ethics perspective taken together with the feminist perspective has showed, virtual child pornographic images flout the sexual mentality based on equality.

In conclusion, from a liberal point of view the prohibition on the production, distribution and possession of child pornography is based on moralism.

Conclusion

First of all I aimed to answer the question whether the harm principle as the 'moral basis' for criminal law does apply to the prohibition on the production, distribution and possession of child pornography. In contrast with the first and the second category of child pornography distinguished, it is difficult to prove harm with reference to the third category of child pornography: virtual child pornography.

Virtual child pornography does not harm children directly. Since no actual children are involved in the production, it does not constitute any victims of sexual abuse. Therefore, the harm principle does not provide for a solid moral basis for prohibition.

The above-mentioned has led to the conclusion virtual child pornography is a 'victimless crime'. According to Bedau, the prohibition of victimless crimes can be based on either paternalism or moralism. Therefore I continued by exploring whether paternalism or moralism could provide for a sufficient moral basis for the prohibition of virtual child pornography.

Paternalism, as an alternative for the harm principle, cannot provide for a moral basis for the prohibition on the production, distribution and possession of virtual child pornography. This is primarily because of the lack of reliable data: it cannot be proven that virtual child pornographic images encourage or seduce children into participating in sexual contacts with adults (which would be required by the *parens patriae* principle), nor that they encourage or seduce paedophiles to commit child abuse (which would be required by legal paternalism).

The broader perspective of virtue ethics taken together with feminism however, shows that it is not necessary to prove a causal link between virtual child pornographic images and actual instances of abuse to consider these images as harmful. Virtual child pornography has turned out to be a 'harmful immorality'. This is because virtual child pornographic images flout the sexual mentality based on the equality norm. With Boutellier, I think the wish to achieve equality in the sexual mentality has led to the will to criminalize the production, distribution and possession of virtual child pornography. From a liberal point of view, the moral basis for prohibition is 'moralism'.

Notes

1. Please see this—and subsequent legal sources—in the "Table of legal documents" following the Reference list.
2. For this and subsequent case references, see "Table of cases" following the references.

References

Bedau, Hugo Adam (1974). Are There Really "Crimes without Victims"? In Edwin M. Schur & Hugo Adam Bedau, *Victimless Crimes / Two Sides of a Controversy* (pp. 55–105). Englewood Cliffs NJ: Prentice-Hall Inc.

Bergman, Sunny (Producer). (2007, March 8). *Beperkt Houdbaar* [Television Broadcast]. Hilversum: Viewpoint Productions & VPRO. Accessible via <http://www.beperkthoudbaar.info>. (Available in Dutch only).

Boutellier, Hans (2000). The Pornographic Context of Sexual Offences: Reflections on the Contemporary Sexual Mentality. *European Journal on Criminal Policy and Research*, *8*, 441–457.

DeAngelis, Tori (2007). Children and the Internet / Web pornography's effect on children. *Monitor on Psychology*, *38 (10)*, 50–52.

Dworkin, Gerald (1972). Paternalism. *The Monist*, *56 (1)*, 64–84.

Feinberg, Joel (1986). Legal Paternalism. In Joel Feinberg, *Harm to Self. The Moral Limits of the Criminal Law* (pp. 3–26). New York: Oxford UP.

Itzin, Catherine (1992). Pornography and the Social Construction of Sexual Inequality. In Catherine Itzin (ed.), *Pornography / Women, Violence and Civil Liberties* (pp. 57–75). New York: Oxford UP.

Johnson, Maureen, & Rogers, Kevin M. (2009). Too Far Down the Yellow Brick Road: Cyber-hysteria and Virtual Porn. *Journal of International Commercial Law and Technology*, *4 (1)*, 72–81.

Kymlicka, Will (2001). *Contemporary Political Philosophy: An Introduction*, Oxford: Oxford UP.

Levy, Neil (2002). Virtual Child Pornography: The Eroticization of Inequality. *Ethics and Information Technology*, *4*, 319–323.

MacIntyre, Alasdair. (1984) *After Virtue: A Study in Moral Theory.* Notre Dame, Indiana: University of Notre Dame Press.

MacKinnon, Catharine A. (1992). Pornography, Civil Rights and Speech. In Catherine Itzin (ed.), *Pornography / Women, Violence and Civil Liberties* (pp. 456–511). New York: Oxford UP.

Mill, John Stuart (1859). On Liberty. Cited in Dworkin, Gerald (1972). Paternalism. *The Monist, 56 (1)*, p. 64.

Moerings, Martin (1999). De verbeten strijd tegen pedoseks en kinderporno. In Martin Moerings, Caroline M. Pelser, & Chrisje H. Brants (eds.), *Morele kwesties in het strafrecht* (pp. 171–193). Deventer: Gouda Quint.

Movisie. (2009). *Seksualisering: "Je denkt dat het normaal is..." / Onderzoek naar de beleving van jongeren.* Utrecht: Felten, Hanneke, Janssens, Kristin, & Brants, Luc.

O'Brien, Matt D., & Webster, Stephen D. (2007). The Construction and Preliminary Validation of the Internet Behaviours and Attitudes Questionnaire (IBAQ). *Sex Abuse, 19*, 237–256.

Quayle, Ethel, & Taylor, Max (2002). Paedophiles, Pornography and the Internet: Assessment Issues. *British Journal of Social Work, 32*, 863–875.

Ruddick, Sara (1975). Better Sex. In R. Baker, & F. Elliston (eds.), *Philosophy and Sex* (pp. 83–104). Buffalo, NY: Prometheus Books.

Rutgers Nisso Groep, Nederlands Jeugdinstituut, Movisie. (2008). *Seksualisering: Reden tot zorg? / Een verkennend onderzoek onder jongeren.* Utrecht: De Graaf, Hanneke, Nikken, Peter, Felten, Hanneke, Janssens, Kristin, & Van Berlo, Willy.

Sunstein, Cass (2001). *Republic.com*, Princeton, NJ: Princeton UP.

United Nations. 1948. Universal Declaration of Human Rights. http://www.un.org/en/documents/udhr/

University of Groningen / Utrecht University. (2004–2005). *Mannen die kinderporno downloaden. Een eerste studie naar inhoudelijke aspecten hiervan.* Utrecht / Groningen: Buschman, Jos.

Table of cases

ECtHR, 2 December 2008, *K.U. v. Finland*, Appl. No. 2872/02. Accessible via <http://www.echr.coe.int/echr>.

Rb. 's-Hertogenbosch, 4 February 2008, LJN: BC3225. Accessible via <http://www.rechtspraak.nl>. (Available in Dutch only)

US Supreme Court, 16 April 2002, *Ashcroft v. Free Speech Coalition*, No. 00-795. Accessible via <http://www.findlaw.com/casecode>.

Table of legal documents

Council of Europe, Convention on Cybercrime and Explanatory Report, Budapest, 23 November 2001 (CETS No.185). Accessible via <http://conventions.coe.int>.

Council of the European Union, Council Framework Decision 2004/68/JHA of 22 December 2003 on combating the sexual exploitation of children and child pornography. Accessible via <http://eur-lex.europa.eu>.

Kamerstukken II 2000/2001, 27 745/3, 4. Wijziging van het Wetboek van Strafrecht, het Wetboek van Strafvordering en de Gemeentewet (partiële wijziging zedelijkheids-wetgeving). Memorie van Toelichting. Accessible via <http://zoek.officielebekendmakingen.nl>. (Available in Dutch only)

UN, Optional Protocol to the UN Convention on the rights of the child, on the sale of children, child prostitution and child pornography. Adopted and opened for signature, ratification and accession by General Assembly resolution A/RES/54/263 of 25 May 2000. Accessible via <http://www2.ohchr.org/english/law>.

UN, Universal Declaration of Human Rights. Adopted and proclaimed by General Assembly resolution 217A (III) of 10 December 1948. Accessible via <http://www2.ohchr.org/english/law>.

US Code, Title 18, Part I, Chapter 110: Sexual Exploitation and other Abuse of Children. Accessible via <http://www.findlaw.com/casecode>.

Wetboek van Strafrecht. Accessible via <http://wetten.overheid.nl>. (Available in Dutch only)

CHAPTER NINE

Virtuality and Trust in Broadened Thinking Online

MAY THORSETH

Introduction—public reason and deliberative democracy

The main issue of this chapter is to show how and in what relevant sense online virtuality might contribute to enhancing public opinion. The enhancement in view is to be understood in line with ideals embedded in deliberative democratic theories of public reason. The problem of the public as raised by Dewey (1927) is my point of departure. Briefly speaking, it is an old philosophical puzzle as to how to develop a better informed public. This need is based in what Dewey identified as an eclipse of the public, and a need to convert the Great Society into a Great Community:

> Till the Great Society is converted into a Great Community, the public will remain in eclipse. Communication alone can create a great community. Our Babel is not one of tongues, but of the signs and symbols without which shared experience is impossible. (Dewey, 1927, p. 142)

Due to political complexity, he recognised on the one hand a need for a better informed public, and, on the other, a need for legislators and policy makers to become better informed of the experiences of the public:

> The essential need ... is the improvement of the methods and conditions of debate, discussion and persuasion. That is *the* problem of the public. (Dewey 1927: 208, quoted in Coleman and Gøtze, 2001, p. 11)

New information technologies of our time partly intend to facilitate resolutions to the problems identified by Dewey and before him by many other philosophers and political thinkers—a problem which is closely related to the ideal of enlightenment found in Kant and many other Western liberal thinkers. How could we ensure that both policy makers and the public are well informed, and what methods are at our disposal?

Before turning to virtual contexts online a few words need to be said on how the problem of the public is interpreted in this paper. The major problem is not the lack of information or a lack of a plurality of arguments and opinions; rather, it is about how to make this plurality serve the function of broadening opinions. Thus, relativism would not be an answer to our problem. Rather, it is about how to retain a plurality of opinions as available to interact in a constructive way, in order to refine a commonly shared argument pool. In Dewey's words it is about "the signs and symbols without which shared experience is impossible" (Dewey, 1927, p. 142). The challenge I want to highlight here is thus how to obtain shared experiences based on a plurality of different outlooks. The sharing of experiences does not necessitate common outlooks, but rather a sharing of signs and symbols. As an example, an adequate understanding of why the ironic cartoons of the prophet Muhammad published in Western media some years ago triggered such a rage among many Muslims would indicate that certain symbols are shared between outraged Muslims and those who are not provoked by the cartoons. Thus, sharing of signs and symbols does not hinder criticism. Rather, the sharing is a prerequisite of it.

Moreover, public reason requires a plurality of autonomous parties. The importance of a plurality of parties is here based in a Kantian understanding of reflective judgment and "the maxim of enlarged thought" (Kant, 1964, cited in O'Neill, 1989, p. 46). Basically, it is about taking account of the possible judgment of others, and it requires the presence of, or potential communication with others. According to Onora O'Neill, "[o]nly communication that conforms to the maxim of enlarged thought can reach the 'world at large'" (O'Neill, 1989, p. 47). O'Neill refers here to the idea that public use of reason has the world at large as its scope—as opposed to private use of reason, which is considered to be more limited in scope, e.g., communication filling the roles of clergy, officers, and civil servants. This definition of the distinction between public and private stems from Kant (1970/1991).

This communicative ideal is also contained in theories of public reason in deliberation. The main purpose has much to do with the ideals underlying deliberative democracy itself, i.e., to improve arguments and opinions through a public test. The communicative ideal of deliberative democracy is very much in line with discourse ethical ideals of unconstrained and free speech found in Habermas (1990), Benhabib (1992), Dryzek (2001), Regh (1997) and many others. Put very briefly, the main idea is that everyone should in principle be able to put forth their opinions and arguments without being obstructed by power relations or other instances of a lack of symmetry between the parties. Here I raise the question whether virtual environments online may serve as a means of weakening the impact of power relations.

The problem of the public today does not differ in kind from the problem described by Dewey. Today's challenge is still one of methods and conditions of debate, discussion and persuasion. However, the increase in amount of available and accessible information appears to create obstacles as well as possible ways of resolving the problem. On the one hand, it is in principle possible to become better informed of the opinions of different publics largely due to the new online conditions of debate. At the same time, and for the same reason, it is also possible for people to pay attention only to the arguments and opinions that attract them. This is the problem identified as filtering by Cass Sunstein (2001). The real problem seems to be how to make people take an interest in rational debate over public issues. In short, civic engagement needs to extend to public as opposed to private use of reason. Arguing from Kant's understanding of "reflective judgment" in the aesthetic domain and Hannah Arendt's extension of it to the political domain (1968), Seyla Benhabib (1992) describes this ability in terms of transcending the subjective conditions of particular judgments. This transcending of subjective and private conditions is about validation of public reason. Unless our judgments address a universal audience we cannot claim universal validity, either.

The validity of reason depends on the judging, and it is not valid for those who do not judge. This point is put forth by Hanna Arendt, in emphasising that the claim to validity presupposes communication between self and others. Hence, a judgment's claim to validity can never extend further than the public realm of those who are members of it (Arendt 1968, p. 221). There is particularly the public aspect of judgment I want to emphasise here, in Onora O'Neill's words, the publicizability of public reason (O'Neill, 1989, p. 33). Any particular judgment is based in contingent and finite appeals that nevertheless may transcend the subjective conditions of the particular judgment. The potential for transcending the purely subjective condition is due to the communicative aspect of all judgments. Hence, reflective judgment is deeply founded in communication. According to Kant himself, we have seen that reflective judgment is supposed to lie outside the political domain, whereas both Arendt and later Seyla Benhabib have argued that it should rightfully be extended to the faculties of politics and morality as well.

Contained in the ideal of deliberative democracy is public use of reason, and a claim to validation by appeal to communication or reflection that extends beyond private, subjective conditions. In John Dryzek's words, the particular has to relate to something universal in order to meet with this claim (Dryzek, 2001, p. 68). In order to transcend the purely subjective conditions of our own situation, I believe it is important to stimulate our creative capacity for imagining potentially different scenarios and judgments from our own, i.e., our capacity for broadened thinking or reflective judgment. One way of doing it is by reflecting on different forms of communication that may be involved in public use of reason. Within the deliberative democracy debate we have seen much discussion of forms of communication, in particular, discussion whether rhetoric, storytelling and the like should be accepted. These forms of communication are conceived in contrast to the calm and dispassionate way of reasoning often associated with discourse ethics. The question whether rhetoric should be accepted has been raised both as a question of rhetoric as a constituent part of all argumentation (Regh, 1997, Young, 2000), and as a means to deliberation (Gutman and Thompson, 2004). I have discussed this elsewhere (Thorseth, 2008).

I have been discussing above public reason against the background of ideals of deliberative democracy without yet questioning different forms of communication. According to some deliberative democrats, rhetoric is viewed as a form of communication that should not be accepted in deliberation, whereas others view rhetoric as a constituent part of all argumentation which can be either criticised or embraced. A third point about rhetoric and passionate communication is whether the online context differs from the offline communication in this respect. I shall move on to examine the virtual aspects of communicative environments online and ask whether emotional or passionate communication seems likely to contribute to public reason in this context. An interesting question is whether there is a possible link between passionate communication and virtuality. An affirmative answer would indicate that the internet is an interesting venue for broadening public reason. Thus, my main interest for now is the communicative potential made possible by the new communicative environments of the internet.

Virtuality of broadened thinking

The main question raised in this chapter is whether online virtuality may contribute to enhancing public opinion, i.e., to get better informed publics. If the virtual can be seen as analogous to something possible—as opposed to the actual—I think there might be a potential for online virtuality to contribute to better informed publics. Before we have a closer look at the very concept of "virtuality," I shall explain how public reason in general, and broadened thinking in particular, calls for it.

Taking my point of departure in Kant's "reflective judgment" or broadened thinking, and the deliberative democratic vision of free and unforced communication, I have emphasised the need to transcend the subjective and private conditions. This is a constitutive constraint in all communication addressing the world at large, i.e., a universal audience.

Including the virtual in shared experiences

The contexts I want to address are situations where our norms and values are put to a test in public space. A concrete example is so-called multicultural conflicts where particular practices are publicly contested and questioned. Typical of many such conflicts is the appeal to particular, culturally embedded norms with a claim to be recognized by an audience that is not itself sharing this norm, be it discussions of wearing religious symbols like the *hijab* in schools, practicing forced marriage or others. To help people extend their imagination beyond the subjective conditions of their own is part of the solution of the problem of the public. The extension beyond private, subjective conditions would further make people share experiences while sustaining a plurality of different opinions in public. One way of overcoming the limitations of one's own conditions is by way of imagining something which is not realised, or to speak with Kant: to put ourselves in the position of everyone else (Kant, 1952, § 40). In Hannah Arendt's words this is "to see things not only from one's own point of view but in the perspective of all those who happen to be present" (Arendt, 1968, p. 221). One means of seeing something from other perspectives than our own is by using our imaginative powers. This happens whenever we read good fiction, e.g., novels—and it also happens when someone tells us a story so vividly that we almost become a part of the story ourselves.

Let me bring an example. A former student of mine told me the story of one of her classmates who was a refugee from Latin America, a woman who had experienced her close relatives being brutally killed before her eyes. She had fled her home country, but she did not belong to a group of refugees who were permitted residence in Norway at that time. Hence, she was not a legitimate refugee. This story made me see that applying the regular standard of admittance to our country was not applicable in this case. The context of the general rules and regulations were challenged by this story, as I could easily see why the particular conditions of my earlier view had to be transcended. Her story induced a reflection about my own view which was contextually based in current laws and regulations. Further, her telling of the story made me realise the contextual limitations of my own initial view of the matter. The story of this woman had an obvious appeal that transcended the particular experiences she had suffered. This is because her telling of the story quite easily made me imagine something virtual which was not actually there, as an aspect of reality—in a sense virtual, but still real!

The following example demonstrates how the contingent facts as seen by someone may be challenged so as to make us imagine something virtual that extends to what is not actually here. Some years ago there was a heated panel debate in Norwegian TV on the increasing tendency among Muslims to have their daughters start wearing the *hijab*. A question was raised whether this change indicated an increase in suppression of Muslim females and a stronger control of Muslim women in Norway. Among the panellists were both Muslims and ethnic Norwegians of both sexes. One of the arguments advanced by a Western feminist ran as follows:

> Muslim women who wear the hijab and live traditional lives are suppressed, and those of you, who disagree, are lying, or else you do not know what you talk about, because you live the lives you do. (TV debate in channel NRK 1 in Norway 2003, on having young girls wear the *hijab* in Norway)

The following counterargument by women wearing the *hijab* was:

> Western women who wear miniskirts and uncover parts of their bellies have no right to criticise us for being suppressed, because you yourselves are not being respected by your own men. (op.cit.)

Both the argument and its counter can be seen as instances of *circumstantial ad hominem arguments*, i.e., arguments that make an appeal to the particular condition of the antagonist in order to pull her down. This undermines the way in which the argument might potentially link the particular circumstances of each party to a universal appeal. In this way they apparently continue to disagree on a universal appeal made by both parties, namely that suppression of females is wrong. Rather than seeing how both parties' circumstances may appeal to a shared vision, none of them succeed in transcending the particular conditions or circumstances. Still, watching this debate contributed to broadened thinking among several of the audience, like myself, as I realised that both parties shared the same vision: to work against female suppression. The vision could be expressed in terms of something virtual, namely a situation of no suppression of females in either cultural context.

Now we may look at this kind of argument in contrast with the claim of democratic deliberation to address a universal audience. The claim is not only to view particular matters from the perspective of everyone else. More importantly, it is a claim to see one's own case from the perspective of others in addressing a universal audience. This is required in order to link particular circumstances to a universal appeal.

Broadened thinking and sensus communis—virtuality defined

This shared vision may be conceived as Kant's *sensus communis* which is a public sense and a critical faculty that takes account of the mode of representation in everyone else (Kant, 1952, § 40). In taking into consideration the mode of representation in everyone else, we are urged to consider not only what is actually real. Additionally, we have to include something virtual as part of the real—or of reality. *Sensus communis* is the power to make judgments for the purpose of public appeal, thereby avoiding the illusion that private and personal conditions are taken for objective:

> This is accomplished by weighing the judgment ... with the ... possible judgments of others, and by putting ourselves in the position of every one else ... [abstracting] ... from the limitations which contingently affect our own estimate (Kant, 1952, § 40, p. 294).

In order to proceed on how online virtuality may be assigned a significant role we shall have a closer look at some interesting views of virtuality in this volume. The chapter in this volume by Johnny Hartz Søraker is helpful in discussing different concepts of virtuality, and I will lean on parts of his contribution when trying to clarify in what sense virtual is interesting to my discussion of public reason, broadened thinking and deliberative democracy. In a standard usage of virtual, something fictional or artificial is brought to mind. In looking at virtuality by way of more philosophically loaded concepts Søraker distinguishes between quite different usages of the term, even if all are related to computers. Not all kinds of virtuality require virtual environments; this is true for communication between multiple users that is computer mediated and interactive while not necessarily happening in three-dimensional environments, like virtual worlds (Søraker, this volume). Thus, we can distinguish between virtual worlds like *Second Life* on the one hand and virtual communities on the other. In the latter there are multiple users and the communication is interactive. This is the kind of virtual community that is often referred to when speaking of online deliberation. To some extent, it makes sense to consider such communicative venues as not necessarily different other than in degree from offline contexts, except for one: in the online context we can communicate intensively with others without knowing who they are. Virtual worlds online that enable a first-person perspective in three dimensional settings, like in *Second Life*, are radically different from the kind of virtual context that I primarily focus on here, and thus it needs to be treated separately in an applied ethical analysis. Just to hint at one major difference: in a virtual and one-dimensional setting the visible appearance seems to be subordinate, whereas visibility obviously is dominant in a three-dimensional setting. Still, I believe both kinds of contexts should be examined further. Here I primarily have in mind the virtual but non-visual context of online communities.

A relevant understanding of virtuality in broadened thinking

Marianne Richter undertakes an analysis of virtuality. According to her, virtuality in the debate on philosophy of virtuality is used both as a modal and as a generic term (Richter, this volume). As a modal term it is used as a contrast to the actual, but as a generic term it is also used as a contrast to the real. For my applied ethical analysis, here it is the modal term which is of uttermost interest, because this understanding makes it possible to see the virtual as distinct from actual or present conditions, while not necessarily viewing it as less real. Hereby an appeal is made to the extension of the understanding of contingencies so as to include transcending of the contingent limitations of the present, or actual. Virtuality is interesting because it captures an important aspect of what may potentially be the case, i.e., what it is possible for us to imagine by getting beyond the contingent actuality we may otherwise be stuck in. For this purpose I believe the online context of virtual communities may contribute more effectively than offline contexts to broadening of public opinion. Let me explain how.

In virtual communities online, the interaction between the participants is among people who do not necessarily know who the others are. In this sense we may view this situation as optimal for deliberative purposes: the better argument is not attached to someone we need to know; in some sense this meets with the discourse on ethical constraints embedded in much of the literature on deliberative democracy. The better argument is not constrained by power relations or other obstacles to free and unconstrained communication. Here we can get some support from the ideals inherent in the theories of Habermas (1990) and Dryzek (2001) on deliberative democracy, but also Kant (1952), Arendt (1968) and Onora O'Neill (1989) in their thoughts of public reason, as discussed above.

There is a huge debate between deliberative and difference democrats about the communicative styles in deliberation, in particular on the importance and relevance of rhetoric, as we have seen above. I have discussed this elsewhere (Thorseth, 2006 and 2008); here I shall confine myself to pointing out one particular way in which the rhetorical, passionate and emotional aspects of communication becomes relevant to virtuality and broadened thinking. The importance of emotions is particularly emphasised by aesthetic democrats who talk about flattening of emotions in online contexts as they view the virtual context online as less immediate and less emotive compared to the real world outside (Marlin-Bennett, 2010). This relates to the democratic attachment that Walt Whitman hoped to engender through his poetry (Frank, 2007, p. 404). The poetic, world-making power associated with aesthetic democracy is the envisaging of people as inexhaustibly sublime. The idea here is that no one can be fully represented; actual presence seems to be a presupposition for democracy. In Marlin-Bennett's understanding of aesthetic democracy, this implies that spontaneous associations and passions are conditions for democracy that are, unfortunately, flattened in online communication. Further, she also expresses pessimism towards a democratic cyber-polis—or virtual *polis*—mainly because it lacks a kind of passionate relationship that embodies a public spirit towards oneself and one's fellow citizens as envisaged by aesthetic democrats. However, reports on how communication is conceived in online contexts obviously varies a lot, not least between savvy (young) and not so savvy (older) generations. Thus, whether passionate deliberation—e.g., by rhetorical and storytelling and emotional modes of argumentation—is flattened in online contexts is yet to be proved.

The emotive and passionate relationship that arises in direct encounters as viewed by aesthetic democrats seems to come very close to the way rhetoric is viewed by difference democrats like Iris Marion Young. Very briefly, Young criticizes deliberative democrats of favouring calm and dispassionate communication, thereby calling attention to "internal exclusion" (Young, 2000, pp. 53–57). This criticism is about disfavouring some people due to lack of effective opportunity to influence others' thinking (Young, 2000, p. 55). The main point to be drawn from this debate here is the disagreement among different kinds of democrats—whether deliberative, difference or aesthetic democrats—as to the role of passions and strong emotions in deliberation. The question of whether online contexts allow for more or less passion as compared to offline reality is most interesting, but has to be left out here to be discussed in a separate paper. One further question to be addressed below is whether distance and lack of embodied encounters may strengthen broadened thinking in deliberation and public reason in virtual communities.

Distance and proximity

In deliberative democracy, the better argument is the ultimate arbiter and highest authority. In order to qualify as such, the argument in view has to win the consent and trust of the audience. What exactly is it that makes some arguments be accepted as trustworthy while others are not? Obviously, they have to be considered truth-preserving in some sense, but the audience also has to trust the speaker, or the medium. This is the point where deliberative, difference and aesthetic democrats seem to diverge. Deliberative democrats are accused by difference democrats of favouring a calm and dispassionate mode of communication (Young, 2000). This is why she claims rhetoric, storytelling and testimony must be added in order to facilitate those who do not manage to express themselves otherwise.

Against deliberative democrats, difference and aesthetic democrats seem to share this view of passions and emotions being constitutive of the better argument, where trust is essential to convincing the interlocutors. By what means an argument is trusted as the better argument seems to be conceived differently among democrats of various kinds.

In discussing modes of communication in deliberation, we can distinguish between two dimensions. On the one hand, there is the passionate/dispassionate dimension; on the other proximity/distance. One might argue that a dispassionate mode of communication serves broadened thinking best. Further, this mode of communication may at first glance appear to fit best with communication in virtual contexts. This seems to be what aesthetic democrats believe, the underlying premise being that virtual, online contexts imply flattening of emotions (Marlin-Bennett, 2010). To some extent, one could then argue that this kind of context would be ideal for broadened thinking. Another premise that seems basic to this line of reasoning is that the virtual context online creates distance between the interlocutors. This does not, however, seem to be confirmed by savvy users of the internet. Rather, proximity and passionate communication seem equally likely to be created due to this context. If passion and proximity are vital to broadened thinking, we have no strong reason to believe that it cannot be established in a virtual, online context. Rather, as virtual environments become ever better at conveying our embodied presence to one another, more trust might be fostered (cf. Ess, this volume).

What seems to be at stake here is the importance, if any, of distance and proximity between the interlocutors. From a deliberative democratic perspective, it could be argued that greater distance, and thus the arguments' independence of who the interlocutors are, will make it easier to address a universal audience, i.e., to make the arguer more attentive to others' perspectives. Against this, aesthetic democrats argue that distance rather promotes flattening of engagement. The upshot of the discussion of distance and proximity, however, seems not yet to be settled in general. First of all, looking further into the virtuality of the online context, it is not clear that there is less proximity when compared to offline communication contexts. Rather, it could be argued that the virtual communities online create a virtual presence among the interlocutors, one that is conceived as even more vivid and passionate than many offline communication communities. If so, then we could expect virtual presence online to contribute to better-informed publics and more broadened thinking.

The distance of the other arguers online may contribute to a better-informed public, but the same could be said of virtual presence and proximity, as well. If online virtuality implies a certain flattening of rhetorical means, i.e., decrease of the role of the speaker's credibility, i.e., *ethos,* and the passions of the audience, i.e., *pathos*, then the criticism from aesthetic democrats does appear convincing, as they claim that that flattening of emotions contributes less to democracy. However, it still remains to be demonstrated whether this criticism is justified.

There is no philosophical knock-down argument to decide whether lack of strong passions will contribute to a better-informed public with a capability for transcending the subjective conditions of their own. If the online communication of virtual communities can be shown to broaden people's thinking, we have no certain basis for suggesting that imaginative powers and engagement are strengthened due to distance rather than proximity. This may be true independently of the question whether strong emotions are set off or not. Strong emotions do not necessarily require embodied meetings if the key issue is the activating of our imaginative powers. Aesthetic democrats have yet to prove that strong emotions do require embodied meetings.

In this section we have looked at virtuality as a possible contribution towards a better-informed public. In the next section we will have a closer look at how trust is another urgent topic for our discussion.

A quest for trust in broadened thinking online

Some of the most interesting approaches to trust that are relevant to the issues raised in this chapter are action-based. According to Jon Elster, trust is "the result of two successive decisions: to engage in interaction and to abstain from monitoring the interaction partner" (Elster, 2007, p. 345). Another Norwegian philosopher, Harald Grimen, joins party with Elster in conceiving of trust as something being displayed in the way people act. But rather than speaking of refraining from taking precautions, or "lowering one's guard" as Elster does, Grimen finds it more adequate to conceive of trust in terms of "keeping the guard down" (Grimen, 2010, p. 195, author's transl.). Both Elster and Grimen share an action-based concept of trust, although there is a difference between lowering and keeping down one's guard. For my purpose here, it suffices to emphasise that trust is displayed in people's choices of action. In an online, virtual context this could range from abstaining from further control of information by someone who looks it up by help of search engines, to dissemination of uncensored information about oneself and by people who participate in virtual communities. Most interesting to our context, however, is the extent to which people are willing to keep the guard down while sharing their opinions with others in virtual communities. The simple fact that people seem to maintain their memberships in virtual communities like, e.g., *Facebook* means that they continue to engage in interaction, and to gradually increase their dissemination of personal data (this is based on several accounts from *Facebook* members). Such willingness is essential to establishing shared experiences.

My concern is how shared experiences in virtual contexts may enhance public reason. A main hypothesis is that more trust paves the way to more broadened thinking in public reason.

There are two aspect of trust I find particularly interesting in the definitions above. One is the action of choosing to trust, or choosing to engage in interaction. A participant in a virtual community who continues her communicative relations is showing trust towards other participants, probably in their capacity as trustees. Lately (spring 2010) there have been clear signs of a phenomenon called "virtual suicide" in different media. This is about people withdrawing from social networking sites such as *Facebook*, thereby committing virtual suicides. Whether this should be interpreted as distrust is not quite clear, as some have pointed to fatigue as the key explanation of it (discussed in the Norwegian radio channel P1, March 2010). Still, it is hard to combat the view that those who do not leave *Facebook* and other virtual communities do keep on engaging in interaction. According to Elster's definition, they make a choice to trust.

Although in a somehow different manner from Elster (2007) and Grimen (2010), John Weckert also argues that trust involves choice (this volume). He relates trust to interpretation of trustees' motives, which makes sense in close relationships. In referring to Rempel et al. (2001), Weckert believes that "[p]eople in high trust relationships interpret their partner's actions in ways that are consistent with their positive expectations" (Weckert, this volume, p. 93). Our choice to trust seems to apply not only to close connections. This is probably true of many relationships in virtual contexts, but also in contexts offline. Harald Grimen mentions an interesting example to illustrate why he thinks that it is not always irrational to choose to trust strangers. On one of his travels he once needed to leave his baggage—two large suitcases and a racing bike—with strangers, while buying his ticket at a railway station in Konstanz am Bodensee. He asked two amiable Swiss ladies in their eighties to look after his baggage (Grimen, 2010, p. 190). The point he makes is that he had no good reason not to trust these old ladies; despite being strangers they were unlikely to do much damage to two big suitcases and a racing bike (Grimen 2010, pp. 192, 194).

People's choice to trust depends on a huge range of factors that obviously differ in various contexts. In comparing situations where strangers are involved, we need to pay attention to the differences between them. Grimen's context is about strangers that we only meet once, as for instance when travelling. The strangers who are engaged in virtual communities are perhaps of a different kind, and hence not necessarily comparable to Grimen's stranger. Still, I think we can apply Grimen's reflection to both kinds of contexts when he argues that it is definitely irrational to trust someone we know to be untrustworthy, while this is not necessarily so only because we know little about them (Grimen, 2010, p. 194). The upshot here is that we may have good reasons to display trust as long as we have no good reason for distrust. Our basis for trust is others' actions. When John Weckert illustrates the point by referring to Kuhn's unquestioned paradigms, he makes a somewhat similar point (Weckert, this volume). Trust, according to Weckert, is usually our "default" position: we keep the guard down as long as we have no good reason for distrust.

When we trust someone, it may be based on previous good experiences of trustworthiness. This makes sense in close relationships in particular, but it may also be extended to strangers by reasoning *ad negativa*: as long as we have no good reason to distrust, we choose to trust. In this volume, Annamaria Carusi also turns to an action-based concept of trust based in a phenomenological analysis of internal relations between perceptions and actions. A main point in her analysis is to conceive of trust as absence of audit evidence. In reflecting on a case of trusting one's babysitter, she suggests that "[t]o trust someone is precisely to underspecify what they are trusted to do or not to do" (Carusi, this volume, p. 111). Underspecifying our expectations the way Carusi discusses seems to come very close to Elster's and Grimen's discussion of refraining from taking precautions, or keeping the guard down. Carusi's understanding of trust adds to the conceptions by Elster, Grimen and Weckert in highlighting the internal as opposed to external relations between and among trustors, where internal refers to "a system of internally related reasons, values and enactments of trust" (Carusi, this volume, p. 115).

Above we have looked at some of the reasons people may have for trust. An action-based concept seems well suited to explaining why we make the choice to trust or not to trust other individuals (or institutions). Starting out from an action-based concept of trust, Bjørn Myskja undertakes an interesting analysis of the connection between trust and benign deception. Trust is thereby being related to the fictional, or virtual. The deceptive aspect of trust is based on an element of fiction in Kant's philosophy of the "as if" (Myskja, this volume). Basically, it has to do with how we relate to the world through hypothetical thinking. The main point is that we think and act *as if* reality were different, a deception that according to Kant is crucial to the arts and aesthetic experience. This *as if* element involved in trust sheds an interesting light on how the virtual—or fictional—may contribute to more broadened thinking, or improvement of public reason. According to Myskja, if we behave as if we trust others—i.e., we pretend to trust—it will generate trust where it does not exist (Myskja, this volume, p. 128). Trust is crucial for our choice of action: if we trust someone, we will act accordingly.

An interesting question is how the benign *as if* deception works between the interlocutors in virtual communities. According to Grimen, we have no good reason not to trust someone unless we have experiences of them acting untrustworthily. The simple fact that we do not know them is no reason for lack of trust. In a similar way, there is no good reason why the *as if* deception should not work in a virtual context. The only difference between offline and online contexts seems to be that there are further aspects of the trustee that may be unknown to us. Several authors have discussed this in terms of a lack of facial expressions and body language we usually make use of for interpreting our interlocutors. This has been discussed in the literature on online friendship (e.g., Nissenbaum 2001, Weckert 2005, and Myskja 2008). However, if we limit ourselves to discussing public reason and broadened thinking, I do not think the lack of bodily presence necessarily needs to play a significant role. Rather, what seems to be more decisive is whether we have had experiences of being deceived by our online interlocutors in a non-benign way. This surely happens sometimes in virtual communities where the purpose of the communication is dating and the like, as when a middle-aged man behaves *as if* he were a teenager. It does not, however, follow that we should not trust, e.g., someone's opinion of the driving forces behind democracy development on the African continent, even if we can't know for sure whether the interlocutor's identity is trustworthy. Thus, we need to be specific about the context to which we apply our theory of trust.

Upshot of argument and concluding remarks

Most interesting to our context is the extent to which people are willing to keep their guard down while sharing their opinions with others in virtual communities. This is because we are interested in how virtual contexts may enhance public reason. A main hypothesis was that more trust paves the way to more broadened thinking in public reason.

Several premises have been discussed, among them, the importance of sharing of signs and symbols. For this purpose I have drawn upon Kant's concept of reflective judgment and *sensus communis* in explaining how to address a universal audience. The broadened thinking in reflective judgment is about including in our judgments scenarios that extend beyond our own private and subjective conditions. For this purpose, different modes of communication, such as rhetoric and passionate communication should not be ruled out. The virtual aspect of judgments should be included in shared experiences. Through a couple of examples I have considered how vivid telling of a story and use of *circumstantial ad hominem* arguments may act as means of including a virtual dimension in shared experiences. The virtual is then conceived as something not actual, but still real. This is based on a modal use of the term. Thus, the virtual is distinct from the actual, but no less real than the present condition being considered. To include the virtual in our judgment of the real means to transcend contingent limitations of the present in our judgments.

In order to share experiences in online contexts we need people to be willing to transcend the contingent limitations of their own experiences. This is how we have described broadened thinking above. Trust is basic to make this happen. This willingness to online trust shows itself in people's choice to engage in communicative interaction. It has been discussed whether imaginative powers are set off by distance or proximity, and, relatedly, by dispassionate or passionate modes of communication. One main issue is whether the online context of our discussion implies greater distance between and flattening of emotions in the interlocutors. As yet, there seems not to be any strong proofs that the absence of bodily presence necessarily means lack of proximity and strong emotions. If so, then there is no reason why the online context should differ in any significant sense from the offline when it comes to deliberation about public matters. Deliberative, difference and aesthetic democracy have served as background for parts of this discussion. Contextualising the discussion is uttermost important in discussing online trust. Obviously, there is a substantial difference between online dating as compared to deliberation about, e.g., democratic development on the African continent. In the latter context, it is of far less importance whether we trust the information of the interlocutor's identity, whereas the opposite holds true in the former.

In this chapter I have looked into virtuality and trust on the background of deliberate democratic ideals. Virtuality and trust are essential to broadened thinking in public reason. On- and offline contexts do not seem to be different in any relevant sense to the kind of discussions that are important to deliberative democracy and broadened thinking. Still, for the purpose of actively transcending peoples' subjective and private conditions the online context may potentially be more powerful in degree. As discussed above, passion and proximity seem to be vital to broadened thinking, and are likely to be created in online contexts, not least as virtual environments are becoming ever better at conveying embodied presence. Still, there is also a possibility that people trust each other in online contexts primarily because they choose communities they already trust, e.g., among academics and intellectuals. So far, I think we can conclude that more empirical data are needed in order to understand better how the connection between trust and broadened thinking work in different virtual environments—a challenging task for another publication.

Acknowledgments

I would like to express my deepest gratitude to Charles Ess for his valuable comments and suggestions for improvement of this chapter, and also for being a very good colleague from whom I have learned a lot through many years. His generosity and sharing attitude has made all our cooperation a great pleasure. Additionally, I also wish to thank Harald Grimen for well-informed and suggestive input to this chapter.

References

Arendt, H. (1968). Crisis in Culture. In *Between Past and Future: Eight Exercises in Political Thought,* (pp. 197–227). New York: Meridian.

Benhabib, S. (1992). *Situating the Self.* Cambridge: Polity Press.

Coleman, S. and Gøtze, J. (2001). *Bowling Together: Online Public Engagement in Policy Deliberation*, Hansard Society. Online at [bowlingtogether.net], 2001 (Downloaded October 2004).

Dewey, J. (1927). *The Public and Its Problems.* New Yoek: Henry Holt & Co.

Dryzek, J. (2001). *Deliberative Democracy and Beyond: Liberals, Critics, Contestations.* Oxford: Oxford UP.

Elster, J. (2007). *Explaining Social Behaviour.* Cambridge: Cambridge UP.

Grimen, H. (2010). Tillit som senket guard [*Trust as lowered guard*]. In R. Slagstad (ed.), *Elster og sirenene* [*Elster and the Sirens*] (pp. 188–200). Oslo: Pax.

Gutman, A. and Thompson, D. (2004). *Why Deliberative Democracy?* Princeton, NJ: Princeton UP.

Habermas, J. (1990). *Moral Consciousness and Communicative Ethics.* Cambridge, MA: MIT Press.

Habermas, J. (1996). *Between Facts and Norms.* Cambridge, MA: M.I.T. Press.

Kant, I. (1952). *The Critique of Judgment*, transl. J.M. Meredith. Oxford: Clarendon.

Kant, I. (1970/1991). *An Answer to the Question: "What Is Enlightenment?"* transl. H.B. Nisbet. In Hans Reiss (ed.), *Kant. Political Writings* (pp. 54–61). Cambridge: Cambridge UP.

Marlin-Bennett, R. (2010). I Hear America Tweeting and Other Themes for a Virtual Polis: Rethinking Democracy in the Global InfoTech Age. In press?

Myskja, B. (2008). The Categorical Imperative and the Ethics of Trust, in *Ethics and Information Technology,* 10 (4), 213–220.

Nissenbaum, H. (2001). Securing Trust Online: Wisdom or Oxymoron? *Boston University Law Review, 81*, 101—131.

O'Neill, O (1989). *Constructions of Reason. Explorations of Kant's Practical Philosophy.* Cambridge: Cambridge UP.

Regh, W. (1997). Reason and Rhetoric in Habermas' Theory of Argumentation. In W. Jost and M.M. Hide (eds.), *Rhetoric and Hermeneutics in Our Time,* (pp. 358–378). New Haven/London: Yale UP.

Sunstein, C. (2001). *republic.com.* Princeton, NJ: Princeton UP.

Thorseth, M. (2006). Worldwide Deliberation and Public Use of Reason Online. *Ethics and Information Technology, 8* (4), 243– 252.

Thorseth, M. (2008). Reflective Judgment and Enlarged Thinking Online. *Ethics and Information Technology, 10* (4), 221–231.

Weckert, J. (2005). Trust in Cyberspace. In R. J. Cavalier (Ed.), *The Impact of the Internet on Our Moral Lives* (pp. 95–117). Albany: State University of New York Press.

Young, I. M. (2000). *Inclusion and Democracy.* Oxford: Oxford UP.

Contributors

Charles Ess is Professor MSO at the Department of Information and Media Studies, Aarhus University (2009–2012). Recent publications include *Digital Media Ethics* (Polity Press, 2009) and, with Mia Consalvo, co-editor, *The Blackwell Handbook of Internet Studies* (2010). With Fay Sudweeks, he co-founded and co-chairs the biennial conference series "Cultural Attitudes towards Technology and Communication" (CATaC).

Annamaria Carusi is Senior Research Associate at the University of Oxford e-Research Centre. Her research interests are in philosophical and social aspects of web-enabled and computational science, social science and humanities. Recent research includes work on models, simulations and visualisations in computational biology and computational social science, images/visualisations/texts across disciplinary boundaries, ethics of e-research, and trust in technology-mediated research infrastructures.

Bjørn Myskja is Associate Professor at the Department of Philosophy, Norwegian University of Science and Technology, Trondheim, Norway. His research interests include Kant's ethics and aesthetics, and the ethics of technology. His most recent work focuses on trust and trustworthiness. He is currently leading two interdisciplinary research projects on nanotechnology and ethics, funded by the Research Council of Norway.

Marianne Richter is a Research Associate at the Institute of Philosophy, University of Stuttgart, Germany. She obtained a Magister's degree in Ethics of Textual Cultures at the University of Erlangen-Nuremberg in 2008, with a thesis on the operationalizability of moral reasoning. At present she is a member of the Cluster of Excellence Simulation Technology (SimTech) and concentrates on methodological and epistemological aspects of visual representation in simulation-based research.

Litska Strikwerda received her LL.M degree in Criminal Law from Utrecht University, the Netherlands, and her MA degree in Applied Ethics from the Norwegian University of Science and Technology (NTNU), Trondheim, Norway. At present she works as an assistant professor at the Willem Pompe Institute for Criminal law and Criminology, Utrecht University, the Netherlands. Besides criminal law, she specializes in the areas of legal and information ethics.

Johnny Hartz Søraker is an Assistant Professor at the Department of Philosophy, University of Twente. His PhD dissertation dealt with the epistemology, ontology and ethics of virtual worlds, with a particular focus on their potential impact on personal well-being. His main interests lies in the intersections between computing and both theoretical and practical philosophy, often related to psychological research. He has published and lectured extensively on issues such as Internet governance, the psychological effects of technology, and the moral status of information.

Mariarosaria Taddeo holds a Marie Curie Fellowship, which is held at the University of Hertfordshire, where she is working on Informational Conflicts and their ethical implications. She is also affiliated with the Information Ethics Group (IEG), University of Oxford. Her primary research interests are philosophy of information, information and computer ethics, philosophy of artificial intelligence and multi-agents systems.

May Thorseth is Professor at the Department of Philosophy, Norwegian University of Science and Technology (NTNU), Trondheim, Norway, director of Programme for Applied Ethics, and also part of the leader group of NTNU's Globalisation Programme. Most of her recent work has focussed on deliberative democracy, in particular related to online communication and virtual environments, and also on democracy and fundamentalism in view of global communication ethics.

John Weckert is a Professorial Fellow at the Centre for Applied Philosophy and Public Ethics and Professor of Computer Ethics in the School of Humanities and

Social Sciences, both at Charles Sturt University. His current research is focused on ethical issues in new technologies, particularly nanotechnology and information and communication technologies. He is editor-in-chief of the Springer journal *NanoEthics: Ethics for Technologies That Converge at the Nanoscale.*

Index

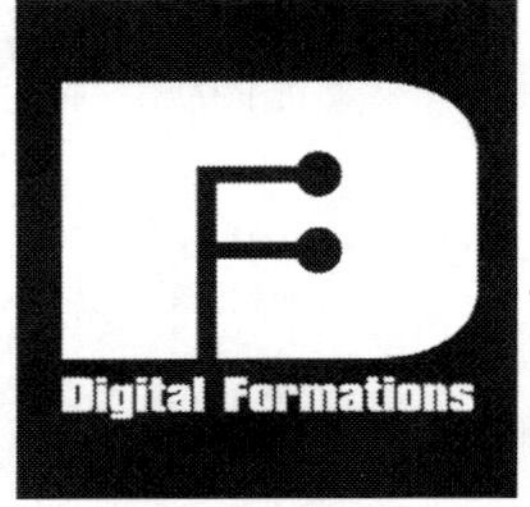

General Editor: *Steve Jones*

Digital Formations is an essential source for critical, high-quality books on digital technologies and modern life. Volumes in the series break new ground by emphasizing multiple methodological and theoretical approaches to deeply probe the formation and reformation of lived experience as it is refracted through digital interaction. Digital Formations pushes forward our understanding of the intersections—and corresponding implications—between the digital technologies and everyday life. The series emphasizes critical studies in the context of emergent and existing digital technologies.

Other recent titles include:

Felicia Wu Song
Virtual Communities: Bowling Alone, Online Together

Edited by Sharon Kleinman
The Culture of Efficiency: Technology in Everyday Life

Edward Lee Lamoureux, Steven L. Baron, & Claire Stewart
Intellectual Property Law and Interactive Media: Free for a Fee

Edited by Adrienne Russell & Nabil Echchaibi
International Blogging: Identity, Politics and Networked Publics

Edited by Don Heider
Living Virtually: Researching New Worlds

Edited by Judith Burnett, Peter Senker & Kathy Walker
The Myths of Technology: Innovation and Inequality

Edited by Knut Lundby
Digital Storytelling, Mediatized Stories: Self-representations in New Media

Theresa M. Senft
Camgirls: Celebrity and Community in the Age of Social Networks

Edited by Chris Paterson & David Domingo
Making Online News: The Ethnography of New Media Production

Zeitfracht Medien GmbH
Ferdinand-Jühlke-Straße 7
99095 Erfurt, Deutschland
produktsicherheit@kolibri360.de